SpringerBriefs in Materials

The SpringerBriefs Series in Materials presents highly relevant, concise monographs on a wide range of topics covering fundamental advances and new applications in the field. Areas of interest include topical information on innovative, structural and functional materials and composites as well as fundamental principles, physical properties, materials theory and design. SpringerBriefs present succinct summaries of cutting-edge research and practical applications across a wide spectrum of fields. Featuring compact volumes of 50 to 125 pages, the series covers a range of content from professional to academic. Typical topics might include

- A timely report of state-of-the art analytical techniques
- A bridge between new research results, as published in journal articles, and a contextual literature review
- A snapshot of a hot or emerging topic
- An in-depth case study or clinical example
- A presentation of core concepts that students must understand in order to make independent contributions

Briefs are characterized by fast, global electronic dissemination, standard publishing contracts, standardized manuscript preparation and formatting guidelines, and expedited production schedules.

More information about this series at https://link.springer.com/bookseries/10111

Khubab Shaker · Yasir Nawab

Lignocellulosic Fibers

Sustainable Biomaterials for Green
Composites

 Springer

Khubab Shaker
Department of Materials
National Textile University
Faisalabad, Pakistan

Yasir Nawab ⓘ
Department of Textile Engineering
National Textile University
Faisalabad, Pakistan

ISSN 2192-1091   ISSN 2192-1105   (electronic)
SpringerBriefs in Materials
ISBN 978-3-030-97412-1   ISBN 978-3-030-97413-8   (eBook)
https://doi.org/10.1007/978-3-030-97413-8

This Springer imprint is published by the registered company Springer Nature Switzerland AG
The registered company address is: Gewerbestrasse 11, 6330 Cham, Switzerland

This work is dedicated to the researchers,

working on sustainable developments,

for greening the world,

into a place to live,

for the future generations.

Preface

The developments in petroleum-based materials and increased dependency on their products have not only resulted into quick depletion of these resources but also raised environmental concerns. The planet earth is under pressure of a huge volume of waste generated as a result of irresponsible production and consumption. A lot of efforts are underway to address these challenges, one of which is the use of sustainable and biodegradable materials. This book is an effort to review the non-conventional lignocellulosic fibers obtained from different plant sources. It also explores the properties of these fibers and their potentials to replace the synthetic fibers as sustainable reinforcement material for Green Composites. Lignocellulosic fibers from leaf, seed, fruit, bast, grass, agrowaste and wood sources are discussed in brief. The chemical composition and properties of these fibers are also compared, to facilitate in material selection. We believe that this book will be essential reading for students and professionals working in the domain of lignocellulosic fibers and green composites.

Faisalabad, Pakistan

Khubab Shaker
Yasir Nawab

Contents

About the Authors

Khubab Shaker obtained his doctorate from National Textile University at Faisalabad, Pakistan in 20018. He is one of the most productive researchers in textiles and composite materials at National Textile University Pakistan, with 60 impact factor publications, 2 edited books, 9 book chapters and more than 25 conference papers and keynote talks. He is currently serving as Assistant Professor and Chairman, Department of Materials at National Textile University, Pakistan. His areas of research include natural fibers, biopolymers, green composites, and their applications as potential replacement of synthetic composite materials.

Address: Department of Materials, National Textile University, Sheikhupura Road, 38000, Faisalabad-Pakistan. e-mail: khubab@ntu.edu.pk

Yasir Nawab is an Associate Professor at the National Textile University, Faisalabad, Pakistan. Currently, he is working as Dean School of Engineering & Technology and leading the Textile Composite Materials Research Group. His research areas are Composite Materials, 2D & 3D woven fabrics and Finite element analysis. He is the author of over 120 peer-reviewed journal articles, 5 books/book chapters and over 50 conference communications. He is the founding Director of National Center for Composite Materials. He received his Ph.D. from the University of Nantes, France in the domain of composite materials in 2012, which was declared as Remarkable Ph.D. in science & Technology in that year. In 2013, he did his postdoc from the ONERA (The French Aerospace lab) and University of Le Havre, France.

Address: Department of Textile Engineering, National Textile University, Sheikhupura Road, 38000, Faisalabad-Pakistan. e-mail: ynawab@ntu.edu.pk

Chapter 1
Green Composite Solutions

Abstract The environmental concerns have forced the material scientists to develop environment friendly materials for various applications. The term "green" is used to describe the materials that are renewable and biodegradable. This chapter covers the different aspects of green composites, including their need, advantages, limitations, research trends, different solutions available and applications. The constituents: fibers and matrix preferred for green composite are discussed briefly. Fabrication techniques commonly used for these composites and their degradation behavior is also explained. The key challenges and applications of these composites are summarized at the end.

Keywords Green composites · Natural fibers · Biopolymers · Fabrication · Biodegradation

1.1 Need for Green Composites

The developments in materials in last centuries have become a destructive force for the globe, leading to consumption of resources at unsustainable rates, producing overwhelming amounts of CO_2 emissions, and adding bulks of toxic waste. It has put a huge burden on the world ecology. The need of time is to develop materials that did little/no harm to the environment and work within the natural capacity of the ecosystem. The green composite materials may be regarded a next generation of sustainable materials, being significantly explored by researchers. These materials include biopolymer-based matrices and natural reinforcements from renewable sources origin [1]. These materials are also alternatively termed as natural fiber reinforced biocomposites, or briefly biocomposites.

The research on green composites is fostered by the environmental issues including pollution and waste management, depletion of petroleum-based resources, carbon emissions, etc. [2]. The term "green" represents the materials that are renewable and biodegradable in nature. It allows for the introduction of these materials as a sustainable, environment-friendly solution of conventional composite materials. The biodegradation and recyclability are the key attribute of green composites; having potential to reduce the environmental impact of plastic pollution. The ancient Egyptians used a composite of clay mixed with straw, and perhaps it may be considered as the first manmade green composite that inspired much more advanced composites. The husk or straw helps to prevent the cracking of bricks, improve its tensile strength, and reduce the erosion rate due to rainfall. These mud bricks have been used to construct houses since thousands of years; and about 30% of the world population still lives in such mud bricks houses.

1.1.1 Advantages of Green Composites

The key advantages of green composites over their synthetic counterparts include:

- Biodegradable and renewable
- Reduced dependency on fossil fuel
- Leading to a circular and low-carbon economy
- Reduced costs, lower density, distinct mechanical properties, and less energy consumption
- Eco-friendliness, formability and cost-effectiveness.

The policy support for bio-based materials across the globe and increasing consumer awareness are considered to be the main driving factors for the growth of green composites market.

1.2   Constituents of a Green Composite

Composite material is a combination of two or more materials differing in composition or form on a macro scale. The composite material exhibits characteristics that are not depicted by any of its components. In case of green composites, the constituents include biopolymer-based matrices and natural reinforcements from renewable sources , as shown in Fig. 1.1. The matrix helps to bind the reinforcement, which is the load bearing component of a composite material. The bio-based resins are derived from starch, vegetable oils, and proteins to replace petroleum-based polymers. While, the lignocellulosic natural fibers, such as jute, flax, hemp, kenaf, sisal, etc. are used as a sustainable reinforcement [3, 4].

1.2.1   Reinforcements

The reinforcements commonly used for green composites are wood and natural fiber based. These fibers are either used as fillers, chopped matt with irregular arrangement, woven or unidirectional architecture. Some researchers have also explored the knitted and 3D structures of natural fibers as reinforcement for green composites. These wood and natural fibers serve as a sustainable source due to their annual renewability and

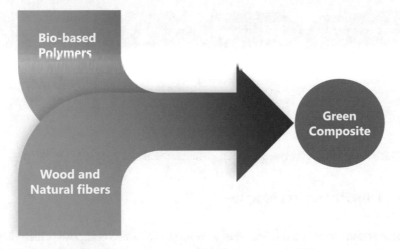

Fig. 1.1 Typical constituents of a green composite material

biodegradation. Additionally, these fibers are low cost, have low density as compared to the synthetic reinforcements and therefore serve as a cheap lightweight solution. The lignocellulosic fibers are preferred for use as reinforcement, as these fibers are sustainable in nature.

1.2.2 Matrices

The polymer matrix systems commonly used for composite materials are broadly categorized into:

- Petroleum based non-biodegradable polymers
- Petroleum based biodegradable polymers
- Bio-based non-biodegradable polymers
- Bio-based biodegradable polymers.

The terms bio-based polymers and biodegradable polymers are often confused, but both are different scientifically. The bio-based plastics are derived from non-petroleum biological resources and may or may not be biodegradable. The biodegradable polymers are degraded via exposure to naturally occurring microbes and may be bio-based or Petro-based. The green polymers are produced partially or entirely from natural resources and are also biodegradable in nature. Generally, the structure of biodegradable polymers consists of ester, amide, or ether bonds [5]. The classification and some common examples of these polymers are shown in Fig. 1.2.

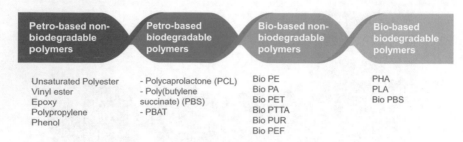

Fig. 1.2 Classification of polymers and some common examples

1.3 Fabrication Techniques

The most commonly used fabrication techniques for green composites include hand layup, compression molding, vacuum infusion and injection molding. Other techniques may also be used keeping in view the material limitations and requirements [6].

1.3.1 Hand Layup

Hand layup is the simplest technique, largely used to fabricate thermosetting polymer-based composites. There may be two variants of this technique, i.e., open molding or close molding. In case of open molding, a surface is used a mold, and mold release agent is applied to this surface for ease of removal. Thin polyester sheets are also sometimes used instead of mold release and serve an additional function of better surface finish. The hardener is added to the resin in prescribed quantity and thoroughly mixed to get homogeneity. A layer of this resin is applied on surface of mold and reinforcement is placed over it. Same procedure is followed again until the desired thickness is achieved. The compaction and void removal are achieved by a cylindrical roller or a brush. It also helps in the uniform distribution of resin in the part. In case of close molding, a second mold is also used to close this part, and then it is allowed to cure for a certain period. Once part is cured, it is removed from the mold for post-curing and further processes.

1.3.2 Compression Molding

The compression molding is a common technique preferred for the fabrication of green composites. It is a close mold technique and can be used to process both thermoplastic and thermosetting matrices. A two-part metal mold is used for this technique, namely the upper and lower mold [7]. Both these parts are matched and metallic in nature. For composite part fabrication, the stack of reinforcement and matrix is prepared and placed between the upper and lower mold at an elevated temperature and pressure. The pressure helps to remove voids from the part, thus helping in part compaction. Once the fabrication is complete, the molds are allowed to cool down and then part is removed. The closed molding helps to get a two-sided finish of the composite part. The process and product are affected by number of parameters including mold pressure, compression rate, heating/cooling rate, curing time, etc.

1.3.3   Injection Molding

It is also a closed-mold technique, suitable for processing a variety of polymeric resins and chopped fibers. In case of direct injection molding, both the chopped fibers and the polymer are fed through hopper into barrel of machine. The polymer is melted in the barrel via application of heat. A rotating screw present in the barrel helps to achieve the uniform dispersion of fibers and the polymer. The screw forces out the mixture of molten polymer and fibers out of the barrel with high pressure, into the mold cavity, through a nozzle. The mold temperature is controlled by cooling/heating arrangement. After cooling down to room temperature, the part is removed from the mold. The schematic is shown in Fig. 1.3c. Sometimes, an additional hopper is also provided for addition of fibers close the end of screw. It helps to avoid the damage to natural fibers due to temperature and shearing action of the screw [8].

Another variant of injection molding is called the extrusion-injection molding, a two-stage process. The first stage involves extrusion of polymer and chopped fiber into a continuous strand, that is cut into pellets by pelletizer. These composite material pellets serve as a raw material for the fabrication of composite materials, using injection molding. The second stage involves the feeding of these pellets into the hopper of injection molding machine [9]. The pellets are converted to molten form in the barrel by heat and forced towards the converging section of barrel, into the mold cavity. The speed of screw, injection pressure and mold cooling rate are the major process parameters that need to be optimized for desired properties.

1.3.4   Resin Infusion/Vacuum Bagging

The resin infusion process is opted for a uniform distribution of resin into the mold cavity. A single sided mold is used for this purpose and the mold cavity is produced by using a bagging film. The reinforcement is stacked on the mold surface and then sealed by a bagging film using sealant tape. Two ports are allowed in this cavity, one for resin inlet and other for the generation of vacuum. The schematic of these fabrication techniques is given in Fig. 1.3d. When vacuum pump is turned on, a vacuum is generated in the mold cavity. The resin flows from resin tank to the

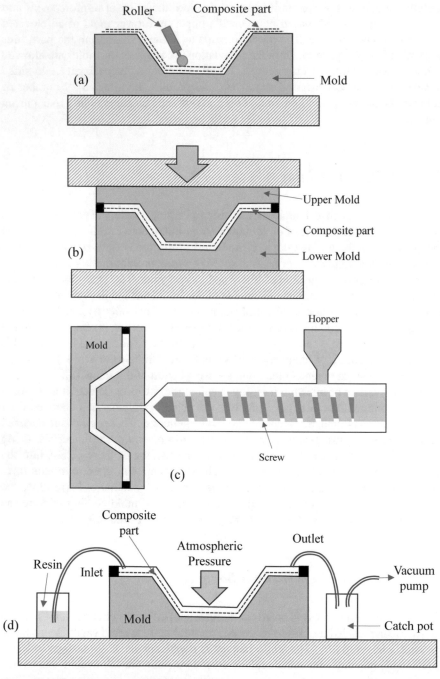

Fig. 1.3 Schematic of composite fabrication techniques, **a** hand layup, **b** compression molding, **c** injection molding and **d** vacuum resin infusion

mold cavity under the action of vacuum and fills the cavity by flowing through and impregnating the reinforcement. The compact and void removal in this approach is ensured by the atmospheric pressure that is applied uniformly over the bagging film [10]. A resin catch pot is attached to the outlet before vacuum pump to collect the overflow resin, avoiding it to flow into vacuum pump.

1.4 Biodegradation of Green Composites

The degradation of any materials is its breakdown occurring through a series of abiotic and biotic processes. The degradation results in a reduction in the molar mass, followed by biological conversion of the breakdown products into carbon dioxide and water. The term biodegradation symbolizes the disintegration of materials by the action of bacteria, fungi, or other biological factors. This disintegration is the result of change in chemical structure, leading to loss of structural and mechanical properties. The metabolic products like water, carbon dioxide, methane, and biomass are released to the environment [5].

Biodegradation occurs in four stages, namely bio-deterioration, bio-fragmentation, bioassimilation, and mineralisation as shown in Fig. 1.4. The bio-deterioration is the process whereby biotic and/or abiotic factors deteriorate the material physically or chemically. Abiotic degradation caused by light, air, heat or other atmospheric factors weakens the polymeric structure, while microorganisms growing on the surface/inside the material are the biotic factors. These microorganisms secrete enzymes that penetrate the polymeric structure, opening channels for influx of air and moisture [11].

The process of depolymerization reduces the molecular weight of polymer chains allowing them to cross the cell membrane. This step occurs under the action of enzymes secreted by microbes. These enzymes act through hydrolytic, oxidative or radicular pathways and can either cause random chain or chain-end scission. Bioassimilation occurs when microorganisms are using the polymeric material as a primary nutrient on which to grow and reproduce.

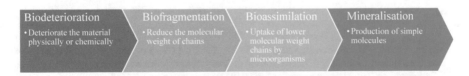

Biodeterioration	Biofragmentation	Bioassimilation	Mineralisation
• Deteriorate the material physically or chemically	• Reduce the molecular weight of chains	• Uptake of lower molecular weight chains by microorganisms	• Production of simple molecules

Fig. 1.4 Stages of biodegradation

1.5   Key Challenges

Although the green composites offer advantages of biodegradation and light weight, but there are certain limitations involved as well. These composites are brittle in nature, exhibiting low impact strength, tend to absorb moisture and have low melt viscosity. These factors affect both the processing and application areas of green composite materials. Some other aspects that need to be considered is the limited availability and higher cost of biodegradable matrices as compared to their synthetic counterparts. For example, the cost of PLA is approximately 2–3 times higher than that of the unsaturated polyester resin. Additionally, the bio-based resins are sometimes developed from edible sources that they may hinder the food supply chain [12]. The further challenges hindering the applications of green composites have been discussed in the Chap. 6.

1.6   Applications

The eco-friendly character of green composites has facilitated their incorporation into various sectors. The countries rich in these resources have the potential of developing this segment for the creation of green economy and strengthening the world's efforts toward sustainability. The global green composites market can be segregated into automotive, aerospace, construction, healthcare, energy, consumer goods, and others [13].

Automotive sector is growing to be a key user of green composites, as the global automotive industry is aiming to produce parts that are lightweight to reduce carbon emissions and enhance fuel efficiency. The current shift is towards the green composites based on plant fibers like flax and hemp, that are not only cheap, but have comparable specific properties. BMW interior designs now use more and more sustainable materials. They aim for sustainable materials, considering the end from the start, involving a long-term thinking and responsible action, and their goal is to achieve a completely climate-neutral business model by 2050.

References

1. M.M. Rehman, M. Zeeshan, K. Shaker, Y. Nawab, Effect of micro-crystalline cellulose particles on mechanical properties of alkaline treated jute fabric reinforced green epoxy composite. Cellulose4 (2019). https://doi.org/10.1007/s10570-019-02679-4
2. K. Shaker, Y. Nawab, M. Jabbar, Bio-composites: eco-friendly substitute of glass fiber composites, in *Handbook of Nanomaterials and Nanocomposites for Energy and Environmental Applications*, ed. by O.V. Kharissova, L.M.T. Martínez, B.I. Kharisov (Springer, 2020)
3. O. Faruk, A.K. Bledzki, H.-P. Fink, M. Sain, Biocomposites reinforced with natural fibers: 2000–2010. Prog. Polym. Sci. **37**(11), 1552–1596 (2012). https://doi.org/10.1016/j.progpolymsci.2012.04.003

4. A. Ashori, A. Nourbakhsh, Bio-based composites from waste agricultural residues. Waste Manag. **30**(4), 680–684 (2010). https://doi.org/10.1016/j.wasman.2009.08.003
5. M. Karnika, K. Shaju, A. Joseph, E.M, Gautheesh, D.A. Gopakumar, S. Thomas, Biodegradation of green polymeric composites materials. Bio Monomers Green Polym. Compos. Mater. 141–159 (2019). https://doi.org/10.1002/9781119301714.ch7
6. M. Pietroluongo, E. Padovano, A. Frache, C. Badini, Mechanical recycling of an end-of-life automotive composite component. Sustain. Mater. Technol.**23**, e00143 (2020). https://doi.org/10.1016/j.susmat.2019.e00143
7. N. AMSC, A. CMPS, Composite materials handbook (2002)
8. F.C. Campbell, *Manufacturing Process for Advanced Composites* (Elsevier, New York, 2004)
9. D.D.L. Chung, *Composite Materials: science and Applications*, 2nd edn. (Springer, London, 2010)
10. Y. Nawab, S. Sapuan, K. Shaker (eds.), *Composite Solutions for Ballistics*, 1st edn. (Cambridge, USA, 2021)
11. M. Warnock, K. Davis, D. Wolf, E. Gbur, Biodegradation of three cellulosic fabrics in soil. Summ. Arkansas Cott. Res. 208–211 (2009)
12. A. Ali et al., Hydrophobic treatment of natural fibers and their composites—A review. J. Ind. Text. **47**(8), 2153–2183 (2018). https://doi.org/10.1177/1528083716654468
13. E. Zini, M. Scandola, Green composites: an overview. Polym. Compos. **32**(12), 1905–1915 (2011). 10.1002/pc

Chapter 2
Lignocellulosic Fiber Structure

Abstract The composition of lignocellulosic fibers varies greatly in terms of cellulose, hemicellulose, and lignin content. The proportions of these components vary with plant fiber specie. The cellulose is a linear macromolecule, and a number of cellulose chains form a microfibril. These microfibrils are arranged at certain angles in the plant cell wall and cemented by hemicellulose and lignin. This chapter discusses the classification of conventional fibers, comparison of their composition, and explains the structure of plant fibers. The chemical composition of lignocellulosic components in also briefly discussed in this chapter. Finally, the novel lignocellulosic fibers explored by researchers from secondary plants are classified, while their details are discussed in subsequent chapters.

Keywords Lignocellulosic fibers · Cellulose · Microfibril · Hemicellulose · Lignin

2.1 Introduction

The fibers obtained from natural resources are classified into plant, animal and mineral based fibers on the basis of origin. The lignocellulosic fibers are the most commonly used alternative of synthetic fibers for composite materials, and these fibers have plant-based origin [1]. Conventionally, these fibers are broadly classified into three categories namely bast fibers, seed fibers, and leaf fibers. Some researchers have also classified plant fibers as primary and secondary, on the basis of plants from which they are obtained. Primary plants are those grown to get fibers, while secondary plants are those producing fibers as a bye-product. Jute, sisal, kenaf, hemp, etc. are the examples of primary plants, while pineapple, coir, oil palm, banana, etc. are the examples of secondary plants [2]. The classification of traditionally available natural fibers is shown in Fig. 2.1.

The textile industry working with apparels and home textiles majorly focusses these natural and synthetic fibers for its raw material. The global worldwide consumption of fibers is about 106 million tons, out of which 62.5% are synthetic fibers and rest are natural fibers. Cotton is the most widely used natural fiber with a share of 25.3%, while other natural fibers account for 4.8% only. Although some researchers have focused to use cotton as reinforcement material for composite but still it is

© The Author(s), under exclusive license to Springer Nature Switzerland AG 2022   11
K. Shaker and Y. Nawab, *Lignocellulosic Fibers*,
SpringerBriefs in Materials, https://doi.org/10.1007/978-3-030-97413-8_2

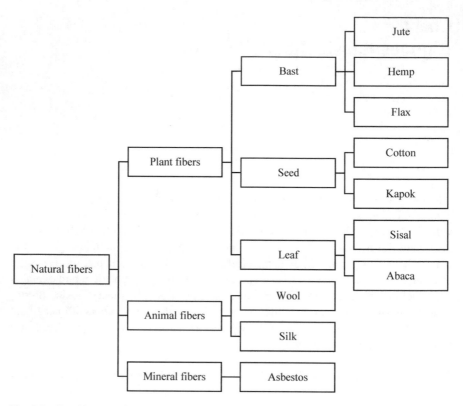

Fig. 2.1 Classification of traditionally available natural fibers [3]

not a preferred choice due to its poor mechanical performance and high moisture content. Therefore, the researchers have focused to use various other bio-resources for extraction of fibers with sufficient mechanical properties.

2.2 Chemical Composition of Lignocellulosic Fibers

The constituents of lignocellulosic fibers include cellulose, hemicelluloses, lignin, ash, etc. The relative fraction of these components varies with the origin of the fiber. The fibrous materials rich in these constituents are also termed as lignocellulosic fibers [4]. Cellulose is the main constituent for all lignocellulosic fibers, contributing about 50–70% of the majority fiber [5]. The constituents of commonly used lignocellulosic fibers are shown in Table 2.1.

It is important to note that the composition of lignocellulosic fibers depends on the method of extraction, the plant growth environment and location, level of maturity when harvested and part of plant from which it is obtained. Hemicellulose has a branched chain structure and is most abundant constituent after cellulose. The lignin

Table 2.1 Composition of commonly used lignocellulosic fibers [6]

Category	Fiber	Cellulose (%)	Hemicellulose (%)	Lignin (%)
Bast fibers	Flax	60–85	14–20.0	1–5
	Hemp	55–90	12–22.4	2–10
	Jute	45–72	13.6–24	9–26
	Ramie	61.8–91	5.3–16.7	0.4–9.1
Leaf fibers	Abaca	53–68	17.5–25	5–15.1
	Sisal	52.8–78	10–19.3	8–14
Seed fibers	Coir	32–43.8	0.15–20	40–45.8
	Cotton	82.7–98	3–5.7	0.75

Table 2.2 Polymeric state and function of natural fiber components

Component	Polymeric state	Function
Cellulose	Highly oriented unbranched molecule, crystalline in nature	Reinforcement
Hemicellulose	Smaller branched molecule, amorphous in nature	Matrix
Lignin	Large 3-D molecule, Amorphous in nature	Matrix

has a highly complex amorphous structure; fills space between cellulose and hemicellulose, thus acting as a cementing material. The polymeric state and function of each component is given in Table 2.2.

2.2.1 Cellulose

Cellulose is a linear chain polysaccharide composed of C, H, and O, and is considered as main constituent for all lignocellulosic fibers [7]. The cellulose chain is composed of repeating β-(1,4) linked glucose units. The two glucose molecules are linked in such a way that the consecutive glucose molecules are rotated 180°. The two 180° glucose molecules together form a cellobiose unit, which is the repeating unit of cellulose (Fig. 2.2), having a generic formula of $(C_6H_{10}O_5)_n$.

Each glucose molecule includes three free hydroxyl (−OH) moieties that can generate hydrogen bonds when interact. These bonds are important in the aggregation of cellulose chains and in determining the cellulose crystal structure. These crystalline chains are organized into two types of microfibrils: cellulose Iα and Iβ. The cellulose Iβ is more stable thermodynamically and dominates in the terrestrial plants [8]. The structure of both cellulose Iα and Iβ is compared in Fig. 2.3.

The properties of plant fiber are determined by the cellulose content of the fibers. The highly hydrophilic nature of lignocellulosic fibers is due to the presence of

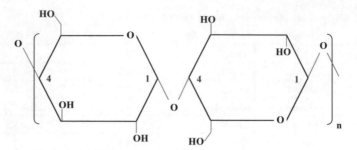

Fig. 2.2 Structure of cellobiose, repeat unit of cellulose

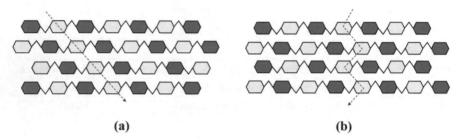

(a)   (b)

Fig. 2.3 Structure of **a** cellulose Iα and **b** cellulose Iβ

hydroxyl groups (–OH) in the cellulose chain. The controlled enzyme or acid hydrolysis of cellulose results loss of amorphous regions, yielding highly crystalline cellulose nano crystals (CNCs). The CNCs are almost defect free, and therefore their modulus is comparable to that of Kevlar and stronger than steel [9].

2.2.2 Hemicellulose

The structure of hemicellulose is complicated, and is made up of several sugar units of different types of pentoses, hexoses, and glycolytic acid. The glycosidic bond exists between the monosaccharide groups in hemicellulose. They often have a backbone made up of a single repeating sugar unit joined by branch points (1–2), (1–3), and/or (1–6). Acetyl and methyl-substituted groups can also be found in hemicellulose. The structure of hemicellulose is given in Fig. 2.4. It acts as cementing material in the plant cell wall, constituting a matrix adjoining cellulosic microfibrils. Contrary to cellulose, the composition of hemicellulose varies with plants. The main differences between cellulose and hemicellulose are given in Table 2.3.

Hemicellulose is less thermally stable as compared to cellulose and lignin, due to its amorphous nature and more availability of hydroxyl groups in the cell wall. It acts like an interfacial coupling agent between highly polar microfibril surface and less

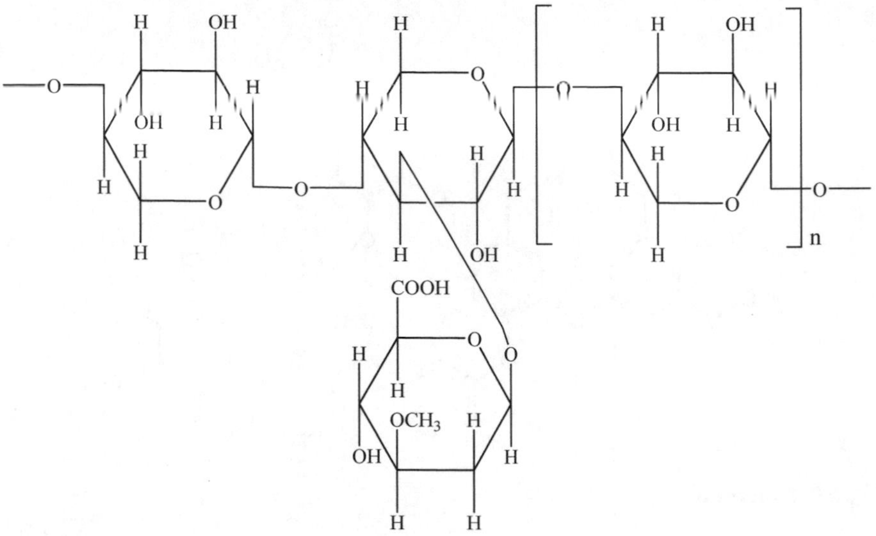

Fig. 2.4 Structure of hemicellulose

Table 2.3 Differences between cellulose and hemicellulose [10]

Cellulose	Hemicellulose
Composition does not vary with plant	Composition varies from plant to plant
Comprises of 1,4-β-d-glucopyranose units only	Contains various dissimilar glucose units
Linear chain	Chain branching is present
Higher degree of polymerisation (10–100 than hemicellulose)	Low degree of polymerisation

polar lignin. A hydrogen bond is formed between hemicellulose and microfibrils, while covalent bond is formed between hemicellulose and lignin [11].

2.2.3 Lignin

Lignin is a polymer of phenyl propane units that are amorphous, extremely complicated, and most aromatic. Lignin is categorized in a variety of ways, although they are most often separated into structural constituents. The three fundamental structure blocks of guaiacyl, syringyl, and p-hydroxyphenyl moieties are found in all plant lignin, while additional aromatic type units can also be found in many different types of plants (Fig. 2.5). Within each plant species, there is a vast range of structures.

Lignin and its associated metabolism play vital roles in the growing of plants. As a building complex phenolic polymer, lignin improves hydrophobic properties

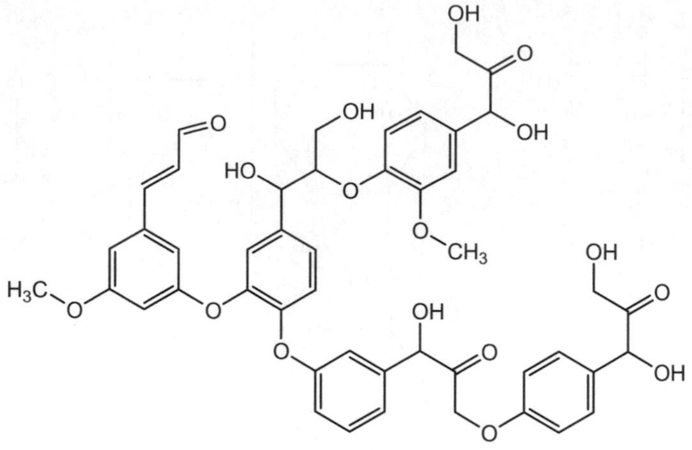

Fig. 2.5 Structure of lignin

by water proofing of cell wall and stimulates mineral transport through the vascular bundle. It also imparts compressive strength and stiffness to the plant cell wall helping them withstand the compressive forces of gravity. Lignin accounts for 25–30 wt% of total plant mass. It is a complex three-dimensional polymer, and its degree of polymerization is ~60, and is amorphous in nature.

2.3   Structure of Lignocellulosic Fibers

The nature has used lignocellulosic fibers as a basic repeat unit of strength-providing skeleton in plants. The fiber bundles are bound together by natural gums, and run through the roots, stems and leaves of plants. These fibers are based upon cellulose, and strands of cellulose fiber are associated with other substances like hemicellulose, lignin, pectin and gums. The cellulose chains are bonded together by hydrogen bonding, that help to form a net-like arrangement called microfibril. A plant fiber is typically a single cell having length and diameter of up to 50 mm, and may be considered as a microscopic tube, i.e., a central cavity (lumen) surrounded by cell wall. The lumen is responsible for the water uptake, while cell wall provides the stiffness to fiber. Plant fibers get their stiffness from cellulose microfibrils that are arranged in lamellae in the plant cell wall around a lumen.

The schematic of plant fiber structure is shown in Fig. 2.6. It can be observed that the cell wall is composed of several rings, segregated into primary and secondary cell walls [12]. Primary cell wall is the outermost layer and is relatively thin and less rigid to allow the expansion during growth. It is composed of a network of orthogonally arranged cellulose microfibril layers. These microfibrils are bonded with the network of crosslinked glycans through hydrogen bonding, providing tensile

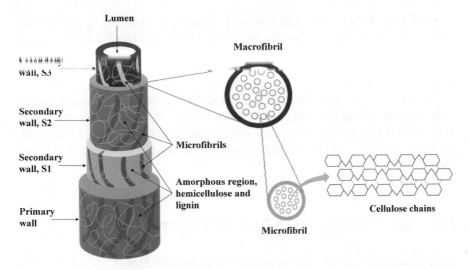

Fig. 2.6   Schematic of plant fiber structure, from cell wall rings to cellulose chains

strength. The pectin network also co-exists in this region and is responsible for compression strength of structure [13]. Both glycans and pectin are heterogeneous groups of branched polysaccharides. The microfibrils typically have a diameter of 10–30 nm and usually 30–100 cellulose chains together form a microfibril [14].

The secondary wall provides rigidity and strength to the plant and consists of three separate layers namely S1, S2 and S3. All these layers are composed of cellulose microfibrils aligned in an orderly manner and this arrangement varies in all three layers. The orientation of microfibrils relative to the axis of cell (fiber) is termed as microfibrillar angle (MFA). In addition to cellulose microfibrils, hemicellulose and lignin are also present in each layer.

S1 is the outermost and thinnest of these layers (thickness range 0.1–0.35 μm) and contributes only 5–10% to the total thickness of cell wall. The MFA in S1 layer is reported to lie in range of 60°–80°. The middle layer (S2) is the thickest of all the secondary wall layers (thickness range 1–10 μm), and most important in determining its mechanical performance, accounting for 75–85% of the cell wall thickness. The MFA in S2 layer is 5°–30° or can be even higher. The S3 layer is innermost secondary wall layer (thickness range 0.5–1.1 μm). The microfibrils are less orderly arranged than in the S2 layer, with MFA 60°–90° [15].

These cell walls differ in terms of composition, and orientation/spiral angle of cellulose microfibrils. As MFA increases, the fibers become less rigid, and longitudinal modulus decreases. The mechanical performance of fiber depends on the cellulose content, degree of polymerization and spiral angle of microfibril [16]. Generally, the fibers having high cellulose content, higher degree of polymerization and a lower microfibril spiral angle show better mechanical properties. Primary walls have a range of cellulose chain lengths (varying from 2000 to 6000) whereas, secondary walls have a more uniform cellulose chain length of about 10,000.

   The lignocellulosic fibers have both amorphous and crystalline regions, and the degree of crystallinity depends on the fiber origin. For example, the cotton, flax, sisal and ramie have high degree of crystallinity (usually 65–70%). A sequential removal of the less organized components results in increased crystallinity. The crystallinity arises partially due to the hydrogen bonding between cellulosic chains, as there are several hydroxyl groups available for interaction. These hydroxyl groups also interact with water via hydrogen bonding and are responsible for the hydrophilic nature of the fiber. All hydroxyl groups present in the amorphous region of the fiber are available to water, while in crystalline region, the water can interact with surface hydroxyl groups only.

2.4   Lignocellulosic Fibers from Secondary Plants

The lignocellulosic fibers from secondary plant sources are categorized on the basis of origin into wood and non-wood fibers, as shown in Fig. 2.7. These fibers are further classified into bast, seed, fruit, grass, leaf and agrowaste based fibers. The agrowaste can be categorized into fibers obtained from husk, straw, bark and yard trimming [17]. These categories of the fibers have been discussed in the Chaps. 3, 4 and 5 of

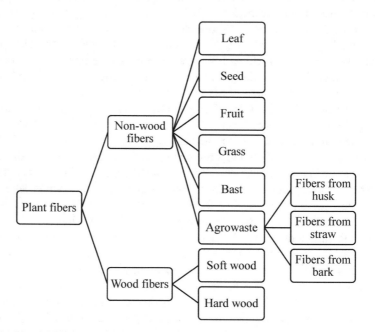

Fig. 2.7 Classification of lignocellulosic fibers from secondary plants

this book. The origin/sources, fiber extraction technique, fiber composition analysis, and properties are discussed and compared.

References

1. C.V. Srinivasa et al., Static bending and impact behaviour of areca fibers composites. Mater. Des. **32**(4), 2469–2475 (2011). https://doi.org/10.1016/j.matdes.2010.11.020
2. O. Faruk, A.K. Bledzki, H.-P. Fink, M. Sain, Biocomposites reinforced with natural fibers: 2000–2010. Prog. Polym. Sci. **37**(11), 1552–1596 (2012). https://doi.org/10.1016/j.progpolym sci.2012.04.003
3. Y. Nawab (ed.), Textile raw materials, in *Textile Engineering* (De Gruyter, 2016)
4. S.M.A. Al kamzari, L.N. Rao, M. Lakavat, S. Gandi, S.P. Reddy, G.K. Sri, Extraction and characterization of cellulose from agricultural waste materials. Mater. Today Proc (xxxx) (2021). https://doi.org/10.1016/j.matpr.2021.07.030
5. S.M. Sapuan, H. Ismail, E.S. Zainudin (eds.), *Natural Fibre Reinforced Vinyl Ester and Vinyl Polymer Composites* (Elsevier, 2018)
6. E. Vázquez-Núñez, A.M. Avecilla-Ramírez, B. Vergara-Porras, M. del R. López-Cuellar, Green composites and their contribution toward sustainability: a review. Polym. Polym. Compos.**29**(9), S1588–S1608 (2021). https://doi.org/10.1177/09673911211009372
7. M.M. Rehman, M. Zeeshan, K. Shaker, Y. Nawab, Effect of micro-crystalline cellulose particles on mechanical properties of alkaline treated jute fabric reinforced green epoxy composite. Cellulose**4** (2019). https://doi.org/10.1007/s10570-019-02679-4
8. M. Sorieul, A. Dickson, S.J. Hill, H. Pearson, Plant fibre: molecular structure and biomechanical properties, of a complex living material, influencing its deconstruction towards a biobased composite. Materials (Basel) **9**(8), 1–36 (2016). https://doi.org/10.3390/ma9080618
9. D.R. Shankaran, *Cellulose Nanocrystals for Health Care Applications* (Elsevier Ltd., 2018)
10. S. Nayak, S. Barik, P.K. Jena, Eco-friendly, bio-degradable and compostable plates from areca leaf, in *Biopolymers and Biocomposites from Agro-Waste for Packaging Applications* (Elsevier, 2021), pp. 127–139
11. D. Jones, C. Brischke (eds.), Wood as bio-based building material, in *Performance of Bio-Based Building Materials* (Elsevier, 2017), pp. 21–96
12. A. Ali et al., Hydrophobic treatment of natural fibers and their composites—A review. J. Ind. Text. **47**(8), 2153–2183 (2018). https://doi.org/10.1177/1528083716654468
13. K. Roberts et al., *Molecular biology of the cell*, 4th edn. (Garland Science, New York, 2002)
14. S. Thomas, S.A. Paul, L.A. Pothan, B. Deepa, *Cellulose Fibers: bio- and Nano-Polymer Composites* (2011)
15. C. Plomion, G. Leprovost, A. Stokes, Wood formation in trees. Plant Physiol. **127**(4), 1513–1523 (2001). https://doi.org/10.1104/pp.010816
16. S. Kalia, B. Kaith, I. Kaur (eds.), *Cellulose Fibers: Bio- and Nano-Polymer Composites.* Springer.
17. K. Shaker, Y. Nawab, M. Jabbar, Bio-composites: eco-friendly substitute of glass fiber composites, in *Handbook of Nanomaterials and Nanocomposites for Energy and Environmental Applications*, ed. by O.V. Kharissova, L.M.T. Martínez, B.I. Kharisov (Springer, 2020)

Chapter 3
Fruit, Seed and Leaf Fibers

Abstract The lignocellulosic fibers are obtained from leaf, fruit, seed, bast, grass, agrowaste and wood. The fibers extracted from leaf, seed and fruit have been discussed in this chapter. Some of the fibers included here are those extracted from Kapok, Brazil nut, Coconut, Borassus fruit, Tamarind fruit, Empty fruit bunch, Agave, New Zealand Flax, Pineapple, Piassava, Palm, etc. The origin, fiber extraction techniques, subsequent processes, cellulose content and some properties of these fibers are detailed in subsequent sections. Majority of these fibers cannot be spun into yarns to produce textiles; however these fibers have been explored by researchers as reinforcement material with different matrix systems. Some details of developed composites are also included in this chapter.

Keywords Fruit fibers · Seed fibers · Leaf fibers

3.1 Introduction

Plant fibers are naturally occurring lignocellulosic fibers extracted from the different plant sources. The non-conventional lignocellulosic fibers are classified on the basis of their origin, as shown in Fig. 3.1. The two main types are non-wood fibers and wood fibers. The non-wood fibers are further categorized into leaf, seed, fruit, bast grass and agrowaste fibers. This chapter is focused on the non-conventional fibers extracted from the leaf, seed or fruit of different plants. The leaf fibers discussed here are agave, New Zealand flax, pineapple, piassava, Leafiran, palm and other fibers. Some examples of fruit and seed fibers included in this chapter are kapok, Brazil nut , coconut, Borassus fruit, tamarind fruit and empty fruit bunch.

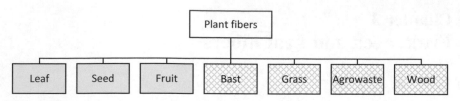

Fig. 3.1 Classification of non-conventional lignocellulosic fibers

3.2 Fruit and Seed Fibers

3.2.1 *Kapok*

The kapok (*Ceiba pentandra*) fibers correspond to the seed hairs of kapok tree. It is a typical cellulose fiber having thin cell wall, large cavity (lumen), low density and shows hydrophobic properties. This fiber has recently received increasing attention of researchers for its full exploration and potential applications [1]. Conventionally, the kapok fibers are used as stuffing for upholstery due to excellent heat insulation, life preservers because of its excellent buoyancy, and sound proofing (insulation against sound). It is also finding usage in advanced applications like composite materials, due to its abundance, renewability, biocompatibility, and biodegradation. It has also been used with other fibers like sisal and glass to produce hybrid composites [2].

Lyu et al. [3] prepared flame-resistant and sound-absorbing composite materials by hot compression technique. They used kapok fibers as reinforcement, polycaprolactone as matrix, and magnesium hydroxide as the flame retardant material. They studied the effect of certain variables including processing temperature, pressing time, composite density, fiber volume fraction, and composite thickness on the properties of composite. The kapok/cassava starch based composites exhibit enhanced breaking stress and modulus [4].

Macedo et al. [5] fabricated composites using untreated and plasma treated kapok fibers with recycled polyethylene matrix. The cold plasma treatment was performed on kapok fibers to enhance the fiber-matrix interfacial adhesion in the composite material. It was found that the storage modulus of PP was increased with the addition of kapok fibers, during dynamic mechanical analysis. The effect was more significant with plasma treated fibers. Kapok fibers also reduced the onset decomposition temperature of developed composites, as determined in thermogravimetric analysis.

3.2.2 *Brazil Nut*

The fruit of Brazil nut tree (Bertholletia excelsa) is in the shape of a spherical capsule, releasing seeds through its lower portion when ripe [6]. The lignocellulosic residue from the Brazil nut can be used for a number of applications, namely ecoplastic filler, biochar production, soil remediation, wastewater treatment, plant substrate preparation, and construction material, as reviewed by Torres [7]. From environmental point of view, the Brazil nut is regarded as a burden free material, as studied in the life cycle assessment study of Brazil nut reinforced starch composites. Bringas et al. [8] prepared starch-Brazil nut fiber composites using three different starch sources as matrix, i.e. corn, sweet potato and Andean potato. They performed the life cycle assessment (LCA) of these biocomposites and found that starch-based green composites were less impacting as compared to the conventional glass reinforced PP composites, and Brazil nut fiber reinforced PLA composites.

Campos et al. [9] prepared biocomposites using Brazil nut pod fiber and high density polyethylene. They used mercerization and electron beam irradiation approaches to improve the mechanical properties of material. The residue of Brazil nut pod was ball milled, and particles less than 125 μm were separated using sieve, and then dried at 80 °C. Mercerisation was done using 20% NaOH solution for one hour. The matrix was pre-irradiated at 15KGy using electron beam accelerator. The composites were fabricated with 20% filler loading in a two-stage process, involving the compounding of fillers and HDPE matrix, and then injection molding into a composite material..

3.2.3 Coconut

The main components of coconut (cocos nucifera) are its fruit, husk and shells. The later are usually disposed-off or burnt considering as a waste. However, these portions are a resource of lignocellulosic fibers which are a potential reinforcement for composite materials [10]. The fiber obtained from coconut are commonly termed as coir fibers, and are used as reinforcement with both thermoplastic (PP, PLA) [11] and thermosetting resins (Epoxy, UP) [12]. The coir is also used in combination with other reinforcement materials like jute, sisal, pineapple leaf fibers [13], oil palm fibers and glass to produce hybrid composite materials. The coconut fibers are treated with caustic soda for surface modification, and it was shown that 10% NaOH concentration results in strong surface modification [14].

3.2.4 Borassus Fruit

Lignocellulosic fibers are extracted from dry ripe fruit of Borassus (palmyra palm) plant. The fruits are spherical in shape and 15–25 cm wide [15]. The fiber extraction is performed using water retting, allowing Borassus fruit to be dipped in water for 14 days. After due time the black skin is removed and underneath it, two types of fibers (short fine and long coarse) are found. The fine fibers adhere to fruit shell while coarse fibers are located edge to edge in fruit nut. These fibers are removed, thoroughly washed in water, and sun dried for a week. The fine fibers are found to have better mechanical properties as compared to the coarse fibers. The average fiber diameter and length were found to be 140 μm and 120 mm respectively. The effect of alkali treatment on the performance properties were also investigated and tensile properties were found to increase due to hemicellulose dissolution.

Sudhakara et al. [16] prepared PP composites reinforced with short fine Borassus fibers using twin screw extruder. They treated the fibers with alkali solution for varying time and found that tensile properties were highest for treatment time of 8 h. Similarly, highest cellulose content was observed for the 8 h alkali treated fibers. A compatibilizer, Maleic Anhydride Grafted Polypropylene (MAPP) was added to the PP for some composite fabrication. It was observed that the mechanical properties of composites having compatibilizer were higher than the untreated and alkali treated fiber composites. When used with epoxy resin, improved tensile properties were observed for resulting composites [17].

3.2.5 Tamarind Fruit Fibers

Tamarind is a leguminous tree bearing edible fruit, consists of fruit, seeds, and pulp; however, the fibers are extracted from fully ripened fruit. The ripe fruits are easily cracked when dry, and fibers, pulp and seeds are separated. The fibers are thoroughly washed to remove any residual pulp and then dried [18]. The thermogravimetric analysis showed that these fibers are thermally stable up to a temperature of 260 °C and are a potential reinforcement for composites that are processed below this temperature [19].

Maheswari et al. [20] developed tamarind fiber reinforced unsaturated polyester composites and investigated their mechanical properties. They concluded that composites with a fiber loading of 25 wt.% exhibit superior mechanical properties. Furthermore, it was also observed that the chemical treatments (alkali and silane treatment) have significantly improved the tensile and flexural properties of the composites, due to improved interfacial adhesion.

Reddy et al. fabricated hybrid composites reinforced with tamarind and glass fiber [21]. They demonstrated that hybrid composites reinforced with fiber length of 2 cm have higher impact strength and hardness as compared to 1 and 3 cm fiber length composites. Nayak et al. [22] extracted tamarind fruit and Borassus fibers and studied the interfacial adhesion strength (IAS) of these fibers with epoxy resin. The untreated fibers showed lower value of IAS as compared to alkali treated fibers.

3.2.6 Empty Fruit Bunch

The empty fruit bunch (EFB) is the waste material produced after the production of palm oil. This EFB is used as source of energy, but high moisture content in it reduces the energy efficiency. The EFB serves as an abundant source of natural fibers

[23]. The fibers are extracted from EFB by mechanical decortication, in which it is subjected to the beating action of the blades. The extraction process is easier when the moisture content in EFB is 35–45%. A higher moisture content results in clogging of the mesh and mill [??]. The EFB was explored as a source cellulose by Ngadi and Lani [24]. They developed a novel approach to extract cellulose from EFB using ionic liquid followed by alkaline treatment. The alkali treatment increased the crystallinity index of cellulose from 38 to 63%. The EFB fibers have been investigated by various researchers as a potential reinforcement for composite materials.

3.3 Leaf

3.3.1 Agave

Agave plants are stemless perennials that can grow in drought-tolerant environments. Several kinds of agave plants contain distinct varieties of strong fibers. The fibers are extracted by mechanical decortication, chemical extraction, or retting, and are strong and stiff, commonly used to make ropes. The fibers are also extracted from the bagasse of agave plant [25]. Some of the most common varieties include Agave sisalana (Sisal fiber), Agave fourcroydes (Henequen fiber) [26], Agave Americana (pita fiber) [27] etc. The fiber obtained from Agave Americana are semi dull in appearance, due to the uneven surface and shape. These fibers have a low density of 1.36 gm/cm^3.

3.3.2 New Zealand Flax

It is an evergreen ornamental plant, native to New Zealand and grows as a clump of long, straplike leaves and yellow or red flowers. This plant also belongs to the Agave family, with stiff leaves that can grow up to 300 cm long and 12.5 cm wide. The scientific name for this plant is Phormium tenax and is also commonly known as harakeke. The harakeke fiber aggregates (also called muka), run along the axis of leaf and are extracted from its upper and lower side [28]. Harris reported the complete process of fiber extraction from these plants [29]. The fibers are separated from the leaf as a continuous strand. After separation, the fibre strand is gently scraped to remove the materials adhering to it. Traditionally the phormium leaves are used in making mats and containers, while fibers are used for weaving and making ropes. These fibers are compounded with plastic waste through screwless extrusion followed by injection moulding for packaging products. It was found that the reinforced plastic properties, tensile strength, modulus and impact strength are significantly increased by the addition of muka fibers [30].

3.3.3   Pineapple

Pineapple is perennial herbaceous plant mainly cultivated for fruit [31], but also produces a huge amount of waste material. The approach most commonly used to extract these fibers is a combination of two techniques, i.e., scrapping and water retting. The scrapping process involves passing the pineapple leaf through three rollers namely feed, scratching, and serrated roller. The layer of leaf is scratched, and leaf is crushed. It allows for the entry of microbes during retting process. These crushed leaves are immersed in water tank having urea or diammonium phosphate for fast retting. Once the retting process is complete, fibers are mechanically segregated, rinsed with water and hang dried in air. The pineapple leaf fibers have a high cellulose content and therefore can serve as reinforcement for green composites [32].

Leao et al. [33] investigated the properties of pineapple leaf fiber reinforced polypropylene composites. The mechanical properties of composites were investigated, and tensile strength and modulus of these composites was 25 and 2800 MPa respectively. The flexural strength and modulus was determined to be 34 and 1900 MPa respectively. Leao et al. [34] also reported the use of PALF for other composites, fiber characterization, modification, composite fabrication and their properties. Some most common matrices used for PALF include epoxy, polyethylene, polypropylene, vinyl ester, unsaturated polyester, polycarbonate and LDPE.

Another variant of pineapple leaf fiber is the curaua leaf fiber. Zah et al. [35] showed that despite being low cost than glass fiber, curaua fibers have physical properties comparable with glass fiber. Tomczak et al. [36] studied the morphology and properties of curua fiber. They showed that an increase in the fiber diameter results in a decrease in the in tensile strength and modulus of the fibers, without any significant change in strain %. Araujo et al. [37] fabricated curaua fiber reinforced HDPE composites with various loading contents, and investigated their mechanical properties. They also used coupling agents to enhance the interfacial adhesion. The composites were fabricated by extrusion and injection molding. The optimum tensile and flexural properties were obtained at 20% fiber loading.

3.3.4   Piassava Fiber

Piassava is a stiff fiber extracted from leaves of Brazilian palm species Attalea funifera. These fibers are upto 4 m long, with an average width of 1.1 m. The extraction is done mechanically by cutting leaf and separating fibers from petiole. A palm tree can yield 8–10 kg of fiber each year [38]. The preliminary results of UD piassava/polyester composites showed that this fiber gives excellent reinforcing effect. In another study, researchers investigated the mechanical properties of Piassava fiber reinforced epoxy composites [39].

3.3.5 Leafiran Fiber

The Leafiran fiber is an additional fiber the leaves of a plant called Typha australis. It is also known as cattail grass and poses a serious problem for agriculture. These are aquatic/semi-aquatic plants with flat blade like leaves, exhibiting excellent mechanical properties with low density [40]. Mortazavi and Moghaddam [41] extracted Leafiran fibers from Typha leaves and investigated their structure, physical and chemical properties. The retted fiber properties depend on the process parameters of the chemical retting. The fibers extracted at 60 °C for 8 h had lower tensile properties than those extracted at 80 °C, due to higher lignin content. The Leafiran fiber consists of cellulose, hemicellulose, lignin, and other materials; the removal of these resulted in separation of fiber bundles in individual form.

3.3.6 Palm Leaf

The palm trees are grown for a variety of economic benefits including oil, fruits, wood and fibers. There are 20 different species of palm leaf fibers available, depending on tree structure and climate. Some of these are discussed here. Atalie and Gideon [42] extracted fibers from Ethiopian palm leaf and characterized the properties. They extracted fibers by mechanical means, water retting and chemical extraction and evaluated their properties. The palm leaf fibers showed strength, elongation and moisture regain comparable to jute, flax and sisal fiber. It was further shown that the water retted fibers show better properties than other two techniques. Al-Oqla and Sapuan [43] explored the potential application of date palm fiber in automotive industry. Comparison between date palm fibers and other fibers showed that these fibers have higher specific modulus and low cost. Sghaier, Zbidi, and Zidi [44] used two approaches (mechanical and chemical) to extract fibers form folioles and leafstalks of doum palm. The fibers extracted from leafstalk exhibited a modulus of 5.99 GPa, tensile strength 154 MPa and breaking elongation of 3.92%. It was further reported that the properties are deteriorated after treatment with caustic soda.

3.3.7 Others

Some other leaf fibers explored by researchers include those extracted from ficus, blue sansevieria and fique leaf. The ficus religiosa leaf fibers are extracted from mature/fallen leaves by water retting for three weeks. The fibers are then extracted by brushing the leaves; rinsed with distilled water and sundried for one week [45]. Finally, the fibers are kept in oven for 24 h at 105–110 °C to remove the moisture. These fibers are also used as a source of cellulose [46]. Sansevieria ehrenbergii, also called blue sansevieria or sword sanseveria, has long, blade-like leaves that serve as a source of ligno-cellulosic fibers. The fibers are extracted using the mechanical decortication process. The fibers can tolerate a peak temperature of 333 °C, and exhibit a tensile strength of 50–585 MPa, with breaking elongation of 2.8–21.7% [47]. Fique plant (Furcrea Macrophylla), also called Cabuya or Maguey is cultivated in Colombia, primarily for the fibers. The fique fiber is extracted from the leaves of this plant and used to make bags. The extraction is performed by mechanical technique [48].

3.4 Composition and Properties of Fibers

The composition (cellulose, hemicellulose and lignin) and properties (density, strength, modulus and elongation, moisture, etc.) of these fibers is compared in Table 3.1.

Table 3.1 Comparative of fibers composition and their properties

Fiber	Composition	Properties	Refs.
Kapok	Cellulose (%) ~53.40 ± 0.23 Hemicellulose (%) ~29.63 ± 0.62 Lignin (%) ~20.73 ± 0.21	Tensile strength ~45–64 MPa Elastic modulus ~1.72–2.48 GPa Elongation ~2–4%	Sangalang [49]
Brazil nut	**Bur fibers** Cellulose (%) ~53 Lignin (%) ~36 Extractives (%) ~7.5 **Shell fibers** Cellulose (%) ~40 ± 0.9 Lignin (%) ~59.9 ± 0.6 Extractives (%) ~5.40 ± 0.06	**Bur fibers** Moisture content (%) ~14.73 ± 0.09 Density (g/cm^3) ~1.398 ± 0.06 **Shell fibers** Moisture content (%) ~10.29 ± 0.03 Density (g/cm^3) ~1.313 ± 0.03	Inamura et al. [6]
Coir	Cellulose (%) ~32–43.8 Hemicellulose (%) ~0.15–20 Lignin (%) ~40–45.8	Tensile strength ~131–343 MPa Elastic modulus ~3.4–17.3 GPa Elongation ~3.7–47%	
Borassus	Cellulose (%) ~82.85 Hemicellulose (%) ~3.02 Lignin (%) ~4.86	Tensile strength ~538–706 MPa Elongation ~31–34% Density ~1.318 g/cm^3	Megalingam et al. [50]

(continued)

Table 3.1 (continued)

Fiber	Composition	Properties	Refs.
Tamarind	Cellulose (%) ~59 Hemicellulose (%) ~22 Lignin (%) ~19	Tensile strength ~61.16 MPa Modulus ~2183.70 MPa Elongation ~6.22%	Maheswari et al. [18]
Empty fruit bunch	Cellulose (%) ~22.5–25.3 Hemicellulose (%) ~24.5–27.8 Lignin (%) ~24–26.6	Tensile strength ~253 MPa Modulus ~16 GPa	Nomanbhay et al. [51], Gunawan et al. [52]
Agave Americana	Hemicellulose (%) ~15 Cellulose (%) ~68–80 Lignin (%) ~5–17	Tenacity ~21–41 cN/tex Modulus ~0.2–1.45 N/tex Elongation ~2–4% Moisture ~8%	Msahli et al. [26], Hulle et al. [27]
Henequen Fibers	Cellulose (%) ~60 Hemicellulose (%) ~25 Lignin (%) ~8	Tenacity ~3.72 cN/tex Modulus ~99 g/den Elongation ~4–6.6%	Msahli et al. [26], Valadez-Gonzalez et al. [53]
Pineapple leaf fiber	Cellulose (%) ~70–82 Lignin (%) ~5–12.7	Tensile strength ~410–1620 MPa Modulus ~32–82 GPa Strain ~1.6–3% Moisture ~11.8%	Asim et al. [32]
Curua fiber	Cellulose (%) ~73.6 Hemicellulose (%) ~9.9 Lignin (%) ~7.5	Tensile strength ~87–310 MPa Modulus ~34–96.1 GPa Elongation ~4.40–4.94%	Souza et al. [54]
Piassava	Cellulose (%) ~48.4 Lignin (%) ~31.6	Tensile strength ~108.5–147.3 MPa Modulus ~1.07–4.59 GPa	d'Almeida et al. [38]
Fique	Cellulose (%) ~63 Lignin (%) ~14.5	Tensile strength ~132–262 MPa Modulus ~3.9–7.5 GPa Elongation ~4.8–10.6%	Teles et al. [48]

3.5 Conclusions

The conventional lignocellulosic fibers (hemp, jute, sisal, etc.) used as reinforcement for green composites are produced in less quantity and have relatively higher cost. The non-conventional fibers can serve as a better replacement for these fibers. These plants are abundantly available, and researchers have explored their potential to serve as a source of fibers. These fibers may be used for conventional textiles in certain cases, but also have a huge potential to serve as a reinforcement material for composites. It will provide the composite designers and engineers a wide range of options in terms of biodegradable reinforcements. Focusing more on the non-conventional sources of natural fibers will help to maintain a cleaner and greener environment, thus contributing to environment sustainability.

References

1. Y. Zheng, J. Wang, Y. Zhu, A. Wang, Research and application of kapok fiber as an absorbing material: a mini review. J. Environ. Sci.**27**(C), 21–32 (2015). https://doi.org/10.1016/J.JES. 2014.09.026
2. G. Venkata Reddy, S. Venkata Naidu, T. Shobha Rani, A study on hardness and flexural properties of kapok/sisal composites. J. Reinf. Plast. Compos.**28**(16), 2035–2044 (2009). https:// doi.org/10.1177/0731684408091682
3. L. Lyu, Y. Tian, J. Lu, X. Xiong, J. Guo, Flame-retardant and sound-absorption properties of composites based on kapok fiber. Materials (Basel) **13**(12), 1–14 (2020). https://doi.org/10. 3390/ma13122845
4. J. Prachayawarakorn, S. Chaiwatyothin, S. Mueangta, A. Hanchana, Effect of jute and kapok fibers on properties of thermoplastic cassava starch composites. Mater. Des. **47**, 309–315 (2013). https://doi.org/10.1016/j.matdes.2012.12.012
5. M.J.P. Macedo, A.L.A. Mattos, T.H.C. Costa, M.C. Feitor, E.N. Ito, J.D.D. Melo, Effect of cold plasma treatment on recycled polyethylene/kapok composites interface adhesion. Compos. Interfaces **26**(10), 871–886 (2019). https://doi.org/10.1080/09276440.2018.1549892
6. P.Y. Inamura, F.H. Kraide, M.J.A. Armelin, M.A. Scapin, E.A.B. Moura, N.L. Mastro, Characterization of Brazil nut fibers. RILEM Bookser. **12**, 71–85 (2016). https://doi.org/10.1007/ 978-94-017-7515-1_6
7. F.G. Torres, K.N. Gonzales, O.P. Troncoso, J. Chávez, G.E. De-la-Torre, Sustainable applications of lignocellulosic residues from the production of Brazil nut in the Peruvian Amazon. Environ. Qual. Manag. 1–10 (2021). https://doi.org/10.1002/tqem.21812
8. P.M. Rojas-Bringas, G.E. De-la-Torre, F.G. Torres, Influence of the source of starch and plasticizers on the environmental burden of starch-Brazil nut fiber biocomposite production: a life cycle assessment approach. Sci. Total Environ.**769**, 144869 (2021). https://doi.org/10.1016/J. SCITOTENV.2020.144869
9. R.D. Campos et al., Effect of mercerization and electron-beam irradiation on mechanical properties of high density polyethylene (HDPE)/Brazil Nut Pod fiber (BNPF) bio-composites, in *Characterization of Minerals, Metals, and Materials 2015* (Springer, Cham, 2015), pp. 637–644
10. A.G. Adeniyi, D.V. Onifade, J.O. Ighalo, A.S. Adeoye, A review of coir fiber reinforced polymer composites. Compos. Part B Eng.**176**, 107305 (2019). https://doi.org/10.1016/j.compositesb. 2019.107305
11. Y. Dong, A. Ghataura, H. Takagi, H.J. Haroosh, A.N. Nakagaito, K.T. Lau, Polylactic acid (PLA) biocomposites reinforced with coir fibres: evaluation of mechanical performance and multifunctional properties. Compos. Part A Appl. Sci. Manuf. **63**, 76–84 (2014). https://doi. org/10.1016/J.COMPOSITESA.2014.04.003
12. P.N.E. Naveen, M. Yasaswi, Experimental analysis of coir-fiber reinforced polymer composite materials. Int. J. Mech. Eng. Rob. Res. **2**(1), 10–18 (2013)
13. F.S. Da Luz, F.J.H.T.V. Ramos, L.F.C. Nascimento, A.B.H.D.S. Figueiredo, S.N. Monteiro, Critical length and interfacial strength of PALF and coir fiber incorporated in epoxy resin matrix. J. Mater. Res. Technol. **7**(4), 528–534 (2018). https://doi.org/10.1016/J.JMRT.2018. 04.025
14. M.A. Esmeraldo et al., Dwarf-green coconut fibers: a versatile natural renewable raw bioresource. Treatment, morphology, and physicochemical properties. BioResources **5**(4), 2478–2501 (2010). https://doi.org/10.15376/biores.5.4.2478-2501
15. K.O. Reddy, C.U. Maheswari, M. Shukla, J.I. Song, A.V. Rajulu, Tensile and structural characterization of alkali treated Borassus fruit fine fibers. Compos. Part B**44**, 433–438 (2013)
16. P. Sudhakara et al., Studies on Borassus fruit fiber and its composites with polypropylene. Compos. Res. **26**(1), 48–53 (2013). https://doi.org/10.7234/kscm.2013.26.1.48
17. P. Sudhakara, P. Kannan, K. Obireddy, A. Varada Rajulu, Flame retardant diglycidylphenylphosphate and diglycidyl ether of bisphenol-A resins containing Borassus fruit fiber composites. J. Mater. Sci. **46**(15), 5176–5183 (2011). https://doi.org/10.1007/S10853-011-5451-6

18. C.U. Maheswari, B.R. Guduri, A.V. Rajulu, Properties of lignocellulose tamarind fruit fibers. J. Appl. Polym. Sci.**110**(4), 1986–1989 (2008). https://doi.org/10.1002/app.27915

19. C. Uma Maheswari, K. Obi Reddy, A. Varada Rajulu, B.R. Guduri, Tensile properties and thermal degradation parameters of tamarind fruit fibers. J. Compos. Mater. **41**(10), 1827–1832 (2008). https://doi.org/10.1177/0731684407087559

20. C.U. Maheswari, K.O. Reddy, E. Muzenda, M. Shukla, A.V. Rajulu, Mechanical properties and chemical resistance of short tamarind fiber/unsaturated polyester composites: influence of fiber modification and fiber content. Int. J. Polym. Anal. Charact. **18**(7), 520–533 (2013). https://doi.org/10.1080/1023666X.2013.816073

21. G. Ramachandra Reddy, M. Ashok Kumar, N. Karthikeyan, S. Mahaboob Basha, Tamarind fruit fiber and glass fiber reinforced polyester composites. Mech. Adv. Mater. Struct.**22**(9), 770–775 (2015). https://doi.org/10.1080/15376494.2013.862330

22. U.A. Kini, S.Y. Nayak, S. Shenoy Heckadka, L.G. Thomas, S.P. Adarsh, S. Gupta, Borassus and tamarind fruit fibers as reinforcement in cashew nut shell liquid-epoxy composites, J. Nat. Fibers**15**(2), 204–218 (2018). https://doi.org/10.1080/15440478.2017.1323697

23. M.S. Sreekala, M.G. Kumaran, S. Thomas, Oil palm fibers: morphology, chemical composition, surface modification, and mechanical properties. J. Appl. Polym. Sci. **66**(5), 821–835 (1997). https://doi.org/10.1002/(SICI)1097-4628(19971031)66:5%3c821::AID-APP2%3e3.0.CO;2-X

24. N. Ngadi, N.S. Lani, Extraction and characterization of cellulose from empty fruit bunch (EFB). J. Teknol.**5**(68), 35–39 (2014). https://doi.org/10.11113/jt.v68.3028

25. S.G. Kestur, T.H.S. Flores-Sahagun, L.P. Dos Santos, J. Dos Santos, I. Mazzaro, A. Mikowski, Characterization of blue agave bagasse fibers of Mexico. Compos. Part A Appl. Sci. Manuf.**45**, 153–161 (2013). https://doi.org/10.1016/j.compositesa.2012.09.001

26. S. Msahli, F. Sakli, J.Y. Drean, Study of textile potential of fibres extracted from Tunisian agave Americana L. Autex Res. J. **6**(1), 9–13 (2006)

27. A. Hulle, P. Kadole, P. Katkar, Agave Americana leaf fibers. Fibers **3**(1), 64–75 (2015). https://doi.org/10.3390/fib3010064

28. I.M. De Rosa, C. Santulli, F. Sarasini, Mechanical and thermal characterization of epoxy composites reinforced with random and quasi-unidirectional untreated Phormium tenax leaf fibers. Mater. Des. **31**(5), 2397–2405 (2010). https://doi.org/10.1016/j.matdes.2009.11.059

29. W. Harris, M.T.U.A. Woodcock-Sharp, Extraction, content, strength, and extension of Phormium variety fibres prepared for traditional Maori weaving. New Zeal. J. Bot. **38**(3), 469–487 (2000). https://doi.org/10.1080/0028825X.2000.9512697

30. K. Jayaraman, R. Halliwell, Harakeke (phormium tenax) fibre-waste plastics blend composites processed by screwless extrusion. Compos. Part B Eng. **40**(7), 645–649 (2009). https://doi.org/10.1016/j.compositesb.2009.04.010

31. M. Asim et al., A review on pineapple leaves fibre and its composites. Int. J. Polym. Sci.**2015** (2015). https://doi.org/10.1155/2015/950567

32. A.R. Mohamed, S.M. Sapuan, M. Shahjahan, A. Khalina, Characterization of pineapple leaf fibers from selected Malaysian cultivars. J. Food Agric. Environ.**7**(1), 235–240 (2009). https://doi.org/10.1016/j.trac.2007.09.004

33. A.L. Leao et al., Pineapple leaf fibers for composites and cellulose. Mol. Cryst. Liq. Cryst.**522**, 36/[336]–41/[341] (2010). https://doi.org/10.1080/15421401003722930

34. A.L. Leão, B.M. Cherian, S. Narine, S.F. Souza, M. Sain, S. Thomas, The use of pineapple leaf fibers (PALFs) as reinforcements in composites, in *Biofiber Reinforcements in Composite Materials* (Elsevier Inc., 2015), pp. 211–235

35. R. Zah, R. Hischier, A.L. Leão, I. Braun, Curauá fibers in the automobile industry—A sustainability assessment. J. Clean. Prod.**15**(11–12), 1032–1040 (2007). https://doi.org/10.1016/j.jclepro.2006.05.036

36. F. Tomczak, K.G. Satyanarayana, T.H.D. Sydenstricker, Studies on lignocellulosic fibers of Brazil: part III—Morphology and properties of Brazilian curauá fibers. Compos. Part A Appl. Sci. Manuf. **38**(10), 2227–2236 (2007). https://doi.org/10.1016/j.compositesa.2007.06.005

37. J.R. Araujo, B. Mano, G.M. Teixeira, M.A.S. Spinacé, M.A. De Paoli, Biomicrofibrilar composites of high density polyethylene reinforced with curauá fibers: mechanical, interfacial and morphological properties. Compos. Sci. Technol.**70**(11), 1637–1644 (2010). https://doi.org/10.1016/j.compscitech.2010.06.006

38. J.R.M. d'Almeida, R.C.M.P. Aquino, S.N. Monteiro, Tensile mechanical properties, morphological aspects and chemical characterization of piassava (Attalea funifera) fibers. Compos. Part A Appl. Sci. Manuf. **37**(9), 1473–1479 (2006). https://doi.org/10.1016/j.compositesa.2005.03.035

39. D.C.O. Nascimento, A.S. Ferreira, S.N. Monteiro, R.C.M.P. Aquino, S.G. Kestur, Studies on the characterization of piassava fibers and their epoxy composites. Compos. Part A Appl. Sci. Manuf. **43**(3), 353–362 (2012). https://doi.org/10.1016/j.compositesa.2011.12.004

40. M.K. Moghaddam, S.M. Mortazavi, Physical and chemical properties of natural fibers extracted from typha Australis leaves. J. Nat. Fibers **13**(3), 353–361 (2016). https://doi.org/10.1080/15440478.2015.1029199

41. S.M. Mortazavi, M.K. Moghaddam, An analysis of structure and properties of a natural cellulosic fiber (Leafiran). Fibers Polym. **11**(6), 877–882 (2010). https://doi.org/10.1007/s12221-010-0877-z

42. D. Atalie, R.K. Gideon, Extraction and characterization of ethiopian palm leaf fibers. Res. J. Text. Appar. **22**(1), 15–25 (2018). https://doi.org/10.1108/RJTA-06-2017-0035

43. F.M. Al-Oqla, S.M. Sapuan, Natural fiber reinforced polymer composites in industrial applications: feasibility of date palm fibers for sustainable automotive industry. J. Clean. Prod.**66**, 347–354 (2014). https://doi.org/10.1016/j.jclepro.2013.10.050

44. S. Sghaier, F. Zbidi, M. Zidi, Characterization of Doum Palm fibers after chemical treatment. Text. Res. J. **79**(12), 1108–1114 (2009). https://doi.org/10.1177/0040517508101623

45. K.O. Reddy, C.U. Maheswari, D.J.P. Reddy, B.R. Guduri, A.V. Rajulu, Properties of ligno-cellulose Ficus religiosa leaf fibers. Int. J. Polym. Technol. **2**(1), 29–36 (2010)

46. K.O. Reddy, C. Uma Maheswari, E. Muzenda, M. Shukla, A.V. Rajulu, Extraction and characterization of cellulose from pretreated ficus (Peepal Tree) leaf fibers. J. Nat. Fibers **13**(1), 54–64 (2016). https://doi.org/10.1080/15440478.2014.984055

47. T.P. Sathishkumar, P. Navaneethakrishnan, S. Shankar, R. Rajasekar, Characterization of new cellulose sansevieria ehrenbergii fibers for polymer composites. Compos. Interfaces **20**(8), 575–593 (2013). https://doi.org/10.1080/15685543.2013.816652

48. M.C.A. Teles, G.R. Altoé, P.A. Netto, H. Colorado, F.M. Margem, S.N. Monteiro, Fique fiber tensile elastic modulus dependence with diameter using the weibull statistical analysis. Mater. Res. **18**(December), 193–199 (2015). https://doi.org/10.1590/1516-1439.364514

49. R.H. Sangalang, Kapok fiber-structure, characteristics and applications: a review. Orient. J. Chem. **37**(3), 513–523 (2021). https://doi.org/10.13005/ojc/370301

50. A. Megalingam, M. Kumar, B. Sriram, K. Jeevanantham, P. Ram Vishnu, Borassus fruit fiber reinforced composite: a review. Mater. Today Proc.**45**(xxxx), 723–728 (2021). https://doi.org/10.1016/j.matpr.2020.02.750

51. S.M. Nomanbhay, R. Hussain, K. Palanisamy, Microwave-assisted alkaline pretreatment and microwave assisted enzymatic saccharification of oil palm empty fruit bunch fiber for enhanced fermentable Sugar Yield. J. Sustain. Bioenergy Syst. **03**(01), 7–17 (2013). https://doi.org/10.4236/jsbs.2013.31002

52. F.E. Gunawan, H. Homma, S.S. Brodjonegoro, A.B. Ba. Hudin, A.B. Zainuddin, Mechanical properties of oil palm empty fruit bunch fiber. J. Solid Mech. Mater. Eng.**3**(7), 943–951 (2009). https://doi.org/10.1299/jmmp.3.943

53. J. Valadez-Gonzalez, R. Cervantes-UC, R. Olayo, P. Herrera-Franco, Chemical modification of henequén fibers with an organosilane coupling agent. Compos. Part B Eng.**30**(3), 321–331 (1999). https://doi.org/10.1016/S1359-8368(98)00055-9

54. S.F. Souza et al., The use of curaua fibers as reinforcements in composites. Biofiber Reinf. Compos. Mater. 700–720 (2015). https://doi.org/10.1533/9781782421276.5.700

Chapter 4
Bast and Grass Fibers

Abstract The lignocellulosic fibers are obtained from leaf, fruit, seed, bast, grass, agrowaste and wood. The novel lignocellulosic fibers of grass and bast origin have been discussed in this chapter. Some of the typical bast and grass fibers included here are those obtained from Isora, Banana, Vakka, Date Palm, Thespesia Lampas, Okra, Bamboo, Bagasse, etc. Most of these plants are not cultivated for fibers, but for food or other products. Fiber extraction is a secondary advantage of these fibers, so they do not disturb the food chain. The origin, extraction techniques, subsequent processes, composition analysis and properties of these fibers are detailed in subsequent sections.

Keywords Bast fibers · Grass fibers · Composition · Properties

4.1 Introduction

Plant fibers are naturally occurring cellulosic fibers extracted from various plant. These fibers are primarily categorized into two types, which are non-wood fibers and wood fibers. The non-wood fibers are further categorized into different types, out of which bast fibers and grass fibers are discussed in this chapter. Ramie, jute, flax, isora, banana, etc. are included in bast fibers category. Bamboo, barley, rice straw, wheat straw, bagasse are included in the grass fibers category [1, 2]. The classification is shown in Fig. 4.1.

The use of these novel lignocellulosic fibers for the fabrication of green composites involves four distinct stages:

(a) extraction of fibers from plant source
(b) characterizing fiber properties
(c) study the interaction between fibers and polymer
(d) investigate performance of green composites.

All these stages have been reported for the different lignocellulosic fibers, reference to their exploration by the researchers.

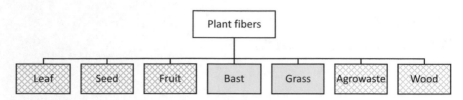

Fig. 4.1 Classification of non-conventional lignocellulosic fibers

4.2   Bast Fibers

4.2.1   Isora

Isora is a lignocellulosic rich fiber extracted from Helicteres isora plant. The fibes present on the inner side of bark are obtained by retting process and may be subsequently treated with NaOH. The SEM analysis showed the fiber diameter to be 10 μm. The fiber had polygonal cross section with oval or circular cavity (lumen). The fiber consists of cells (crystalline cellulose) that are embedded into the matrix (non-crystalline hemicellulose-lignin) [3].

The chemical treatments affect the properties of fiber including its composition, thermal stability, crystallinity, etc. The crystallinity index of raw fiber is 34.11, which is increased by alkali treatment, H_2O_2 bleaching and acid hydrolysis to 49.26, 77.25 and 88.68 respectively. These treatments also enhance the thermal stability of the fiber. The TGA curve shows an initial weight loss between 60 and 110 °C due to the vaporization of water, while weight reduction began at 220 °C due to the degradation of hemicellulose. Massive degradation observed in the temperature region from 250 to 450 °C due to degradation of lignin and cellulose.

Isora nanofibers (INF) are extracted using a combined thermal-chemical–mechanical approach [4], as shown in Fig. 4.2. Joy et al. [5] used INF as reinforcement to fabricate PBS (Polybutylene succinate) based nanocomposites. The fabrication of nanocomposites involves melt mixing of PBS and INF in a twin-screw compounder, extrusion into pellets and injection molding into desired part. The addition of INF

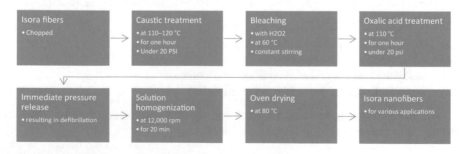

Fig. 4.2 Process flow for the extraction of isora nanofibers (INF)

leads to enhanced tensile and flexural moduli of the composites but was accompanied with a reduced toughness and breaking elongation. The optimum fiber loading for enhanced properties was found to be 1.5 phr (parts per hundred parts of polymer resin).

4.2.2 Banana

Banana (Musa sepientum) is a fruit producing plant having more than 300 species, with annual production of approximately 70 million metric tons. The plant fruits only once in its life and therefore is cut for new plant to grow. These cut pseudo stems are burnt off, but they are a good source of lignocellulosic fibers having relatively good mechanical properties [6]. The "pseudo-stem" represents a clustered, rolled bases of aggregated leaves.

The mechanical decortication is one of the most commonly used methods for extraction of banana fibers from the stem. The longitudinal slices are prepared from stem and fed to mechanical decorticator, where mechanical action leads to the separation of pulp and fibers. The extracted fibers are air dried in shade. Ebisike et al. extracted [7] fibers from banana stem by chemical retting technique under four different concentrations of NaOH. The highest yield % (0.55) was observed for 0.0 M NaOH concentration. They also found that the optimum pH for chemical retting of banana is in the neutral pH region (pH = 7).

The banana fiber has a complex structure, having helically wound microfibrils in amorphous matrix of hemicellulose and lignin. The diameter of banana fiber varies from 80 to 250 μm. Murali and Mohan [8] showed that banana fiber has nearly circular cross-section with an area of 0.3596 mm^2. Guimaraes et al. [9] studied the processing and characterization of banana fiber reinforced corn starch composites. The extracted fibers were shredded, air-dried for one week and milled in a ball mill for upto three hours. The ball milling produced banana fibers having average length 3–5 mm. Starch was used as matrix material, and glycerin was added to it as a plasticizer. The fibers and starch were homogenized in a ball milling machine. Subsequently, mixing was performed in a vibratory mill for 1–3 h. Finally, the mixture was poured in molds and subjected to pressure and temperature.

The banana fibers are also being incorporated in different thermoset matrices, as reported by different researchers [10]. The banana fiber reinforced polyester composites showed a specific modulus of 2.39 GPa, which is of the same order as that of glass fiber composites.

4.2.3 Vakka

The vakka plant, also called as royal palm (Roystonea regia) is abundant in nature. The fibers are extracted from the foliage (leaves and stem sheath) of the plant which sheds on to ground. Water retting is recommended for the extraction of fibers, as it is the most economical and simple method. The sheath is separated from leaves, dried in shade and immersed in water for 15 days. The sheath contains layers of fibers throughout its thickness. The outer most layers on the sheath loosen at the end of 15 days and are separated, washed and immersed in water for three more days. Subsequently it is rubbed and rinsed with water to get the fibers [8].

The fiber growth in the sheath is in longitudinal direction, producing a continuous fiber of 1000–1500 mm length. The fibers obtained are white in color, with a relatively rough surface that expectedly enhances the adhesion with matrix in composites. Therefore, these fibers have the potential to be a good reinforcement for composites, having an enhanced advantage of being lightweight.

For composite fabrication, the fibers can be used in unidirectional form, without being converted into fabric. Murali [11] and Ramanaiah [12] used unsaturated polyester resin to produce the vakka fiber reinforced composites. Both the studies focused on the investigation of thermal and mechanical properties of composites. Murali showed that tensile strength and modulus of vakka fiber composites is much higher than those of sisal and banana fiber composites with fiber loading of 40%. Ramanaiah reported that thermal conductivity of vakka fiber composites decreases with increase in fiber loading (0.179 W/m K at 0.4 fiber fraction).

4.2.4 Date Palm

The date palm (Phoenix Dactylifera L.) plant is abundantly produced around the globe (more than 4000,000 ton). The fibers may be obtained from four different parts of the date palm, i.e., bark or sheath, spadix stem, petiole (midribs) and leaflet. Several techniques have been used to extract fibers from these different plant parts, including mechanical, chemical and biological extraction [13]. Mechanical extraction is mainly used for the leaflets, where these are shredded into thin slices and used as reinforcement. The fibers from dry sheath are also extracted by mechanical extraction, due to its open form. The chemical extraction is equally used to extract fibers from all parts of plants. The biological retting has been used to extract fibers from the mesh, spadix stems and leaflets.

The diameter, length and properties of fibers vary greatly according to the parts of plant and extraction technique. The diameter of fibers extracted by biological retting from spadix stem ranges from 20.3 to 20.6 μm, while its length is 0.56–0.74 mm. The thermal stability of chemically extracted fibers is higher than that of the mechanically extracted fibers. This is possibly due to the removal of hemicellulose due to chemical treatment that degrades at lower temperature [14]. The fibers degrade in three stages; losing moisture, then hemicellulose and cellulose and finally lignin are last to degrade.

Owing to its better thermomechanical properties, the date palm fibers are used for conventional textile applications, as well as for composite material reinforcement. The researchers have used date palm fibers with different matrices like thermoset, thermoplastic and biopolymers, using compression molding, extrusion and injection techniques. Generally, it has been reported that using date palm reinforcement with different polymers has the potential of producing composite materials with enhanced performance, in terms of tensile, flexural, thermal and acoustical properties [15]. Owing to this, these composites may be used for a number of applications, especially in automotive.

4.2.5   Thespesia Lampas

Thespesia lampas is an evergreen shrub (2–3 m tall), and has medicinal uses, while its stem produces lignocellulosic fibers. For fiber extraction, the cut stems are immersed in water for 24 h to separate the bark from the stem. Afterwards, bark is removed and fibers are extracted from bark with combined mechanical and water retting process. The separated fibers are rinsed with water and dried to remove moisture. This process gives a fiber yield of about 60% [16].

Alkali treatment of these fibers with NaOH and subsequent neutralization with acetic acid helps to remove impurities and hemicellulose. Chemical properties, morphological, structural, and tensile properties are changed by alkali treatment of these fibers. The diameter of untreated fiber was 64 μm, while that of alkali treated fibers was 61 μm. Scanning electron microscopy of fiber surface shows that untreated fibers have surface impurities like wax and fatty substances. These impurities are removed by alkali treatment; also causing fibrillation due to removal of binder. The tensile properties and crystallinity were increased with alkali treatment. This increase in the mechanical properties and crystallinity index is due to removal of hemicellulose. The TGA thermogram showed that onset degradation temperature increased from 230 to 270 °C for alkali-treated fibers [16].

A comprehensive analysis of the properties shows that the fibers have superior mechanical performance as compared to conventional plant fibers. The fiber is thermally stable even upto 230 °C. All these factors show that the alkali-treated fibers may be used as reinforcement, even with high melting thermoplastic polymers. These fibers may also be used in applications such as paper pulp or as a source of cellulose.

4.2.6   Okra Fibres

Okra bahamia (Abelmoschus esculentus) is a vegetable plant most commonly grown in tropical and warm temperate regions. Okra is an inexpensive source of carbohydrates, protein, vitamins, and other nutrients [17]. The lignocellulosic fibers are extracted from bark of this plant. This plant produces a huge amount of natural fibers, as stems are thrown in the field after collecting vegetable. For fiber extraction, the cut stems of okra plant are immersed in water for one month to remove gummy and cementing material. The fibers are then separated from the stems and washed with water three times. These fibers are then tied with ropes, air-dried, and stored in moisture-proof containers [18].

Microscopic examinations of the fiber shows that it has a polygonal cross section that varies considerably from circular shape. The average fiber diameter was reported to be 88.34 ± 27.33 μm. The central cavity (lumen) size is 0.1–20 μm, and the cell wall is 1–10 μm thick [18].

The TGA thermogram of okra fiber shows three stages of weight loss: initial 8% loss between 30 and 110 °C, onset of degradation at 220 °C and loss of 16.1% between 220 and 310 °C and final weight loss of 60.6% between 310 and 390 °C. The weight loss in first step is attributed to the vaporization of the water, while second and third step are attributed to the depolymerization of pectin, hemicellulose and glycosidic linkages and the degradation of α-cellulose respectively.

The surface properties of okra fibers are modified by bleaching, alkalization and grafting, maintaining its inherent mechanical properties. The fibers obtained after grafting showed better physico-chemical properties and may serve as potentially compatible reinforcement with several resins for composites. The biodegradation, mechanical properties, better thermal properties and moisture resistance are the other parameters augmenting its use as reinforcement material for various applications [19].

4.2.7 Hibiscus Sabdariffa (Roselle)

Hibiscus sabdariffa is a flowering plant and can survive in different climatic conditions with little care. It is a non-conventional source of bast-fibers, and traditionally people make ropes using its fibers for domestic applications. A number of approaches have been explored for fiber extraction including water retting, enzymatic and microbial retting of decorticated fiber. Kalita et al. [20] used a combination of mechanical decortication and chemical treatment to extract fibers from stem of Hibiscus sabdariffa. For degumming, bundle of decorticated fibers was immersed in a bath containing sodium carbonate solution for a certain period. The degummed fibers were thoroughly washed, neutralized and dried at ambient temperature. They further bleached the fibers and applied a softening agent to make them suitable for spinning into yarn.

The average bleached fiber length was 118 mm, while its diameter was found to be 11.0 μm. The yarn was spun with a twist per inch level of 18, using both degummed and bleached fibers. The tenacity of degummed yarn was higher (17.2 g/tex) as compared to bleached yarn (15.86 g/tex), while breaking elongation was high for bleached yarn (11.9%) as compared to degummed yarn (10.8%). The spinning of fibers into yarns allows the production of fabrics, that may find application in textiles, as well as serve as reinforcement material for composites.

Singha et al. [20] modified the Hibiscus fibers using silane coupling agents and evaluated their properties. They showed that silane treatment has improved the physicochemical properties of Hibiscus fibers and can be a suitable alternative to the synthetic fibers. Singha and Thakur [21] developed composites using Hibiscus fibers as reinforcement with phenol formaldehyde as matrix material. The fibers were chopped to 3 mm length and mixed in different fractions (10, 20, 30 and 40%) with resin. The mixture was spread uniformly on mold surface to fabricate composite materials. Highest value of strength in tensile and compression test was observed for 30% fiber loading (31.8 MPa and 99.8 MPa respectively). The flexural strength and modulus were reported to be 457 MPa and 11.75GPa respectively at 30% fiber loading.

4.3 Grass Fibers

4.3.1 Bamboo

Bamboo belongs to the grass family Bambusoideae and is one of the fastest growing plants in the world. Some researchers include it in the bast fibers but basically it belongs to the grass family and may be termed as a grass fiber. It has a unique structure, where cellulose fibers are aligned along the culm, embedded in a lignin matrix [22]. The percentage of fibers increases from bottom towards the top of the culm.

The bamboo fibers are extracted by combination of mechanical and chemical methods [23]. The bamboo is cut into strips (width 1.5–1.75 cm, thickness 0.65–0.75 mm). These strips are soaked in 0.1 N NaOH solution for 72 h to ease in fiber separation. The strips were washed and dried at room temperature and subjected to mechanical action for fiber extraction. A high temperature and NaOH concentration promote the dissolution of hemicellulose, but cellulose is also damaged when NaOH concentration is above 0.45 N at 110 or 120 °C [24].

Several forms of bamboo may be used as reinforcements, like fibers, strips, sections or even the whole bamboo [25]. Bamboo fibers can be used in synthetic polymer composites such as in unsaturated polyester and vinyl ester composites [23], as well as green composites.

4.3.2 Bagasse

The sugarcane refers to several species of tall perennial grass in the genus Saccharum, tribe Andropogoneae. The sugarcane plants are 2–6 m high, rich in sucrose and are used for sugar production. The waste fibrous residue of sugarcane after extraction of juice is termed as bagasse and used as a source of lignocellulosic fibers. These fibers are not included in the category of bast fibers, that represents fibers obtained from the stem of plant. In bast fibers, the arrangement of fiber bundles is regular and in a definite ring pattern, while in sugarcane the fibers are randomly dispersed [26].

The fibers are obtained from the bagasse using kraft process, which use 17% NaOH solution to extract the cellulosic fibers. Sometimes, prehydrolysis is performed in an autoclave at 130 °C for one hour as a pretreatment. Finally, the fibers are neutralized by a series of washing processes, to eliminate the excess NaOH in the fibers. The fibers are then oven-dried at 105 °C for 24 h, and conditioned under standard atmospheric conditions [27].

A huge amount of lignin can be extracted from the sugarcane bagasse fiber with organosolv process. This lignin content is used with phenolic type resin from 40 (lgnin-phenol–formaldehyde) to 100 wt% (lignin-formaldehyde) substitution. The consequences gotten are assuring for the use of sugarcane bagasse as raw materials for making fiberboards to be used in hot areas [28]. Luz et al. [29] used three variants of bagasse (sugarcane bagasse, bagasse cellulose and benzylated bagasse) to reinfroce polypropylene. The composites were fabricated using injection molding and compression molding at 205 °C with a hearing rate of 2 °C/min for 3 h. The tensile and flexural strength was not good due to non-homogeneity of fibers and blisters. Sun et al. [30] used bagasse as a source for the extraction of hemicellulose by ultrasonic treatment.

4.3.3 Other Grass Fibers

The Alfa fibers are extracted from esparto grass (Stipa tenacissima) leaves using alkaline treatment to remove the non-cellulosic substances such as pectin, lignin, and hemicelluloses. Alfa fibers have special cellulosic nanoparticle and chemical composition, which enable them to use as reinforcing material for nanocomposites. Alfa reinforced nanocomposite, these fibers have polysaccharide and lignin content about 70 and 20% wt., respectively. Alfa fibers produce two types of nano-sized particles, one is nanocrystal cellulose and another one is microfibril cellulose. An aqueous dispersion of cellulose nanoparticles with acrylic latex can be casting for nanocomposite manufacturing. This will enhance stiffness over the glass transition of the matrix [31, 32].

Mendong fibers are bio fibers, which extracted from Mendong Grass (Fimbristylis globulosa). These fibers are extracted mechanically and then immersed in water for a week. Finally, the fibers are cleaned, washed, dried and subsequently treated with alkali solution. The composition, and properties of Mendong fibers indicate that it is suitable for fibrous applications after alkali treatment. The mechanical properties render it a suitable option for manufacturing lightweight composite materials [33].

Napier grass fibers are those special fibers which belongs to the Poacease family of Pennisetum purpureum class. This requires very little nutrients for its growth. Reaping period for the Napier grass fiber is 3–4 months with a dry biomass yield. The fibers are extracted by conventional retting process. The fibers were washed thoroughly in distilled water and dried. Caustic treatment is performed to remove the hemicellulose and waxes. Finally, the fibers are neutralized, cleaned and dried. Characterization of fibers is done with a scanning electron microscope, FTIR, NMR method. The thin and rough surfaces of these fibers enhance tensile properties. This analysis illustrations the feasibility of employing the by-product Na-pier grass fibers as reinforcing material for composites [34, 35]

4.4 Composition and Properties of Fibers

The composition (cellulose, hemicellulose and lignin) and properties (density, strength, modulus and elongation) of these fibers is compared in Table 4.1.

Table 4.1 Comparative of fibers composition and their properties

Fiber	Composition	Properties	Refs.
Isora	Cellulose (%) ~72.0 ± 3.1 Hemicellulose (%) ~3.1 ± 2.4 Lignin (%) ~21.2 ± 0.9	Tensile strength ~500 MPa Elastic modulus ~26.3 GPa Elongation ~2.5%	Chirayil et al. [4]
Banana	Cellulose (%) ~31.3 ± 3.6 Hemicellulose (%) ~14.9 ± 2.1 Lignin (%) ~15.1 ± 0.7	Density ~1350 kg/m^3 Tensile strength ~54–754 MPa Elastic modulus ~7.7–20 GPa Elongation ~10.35%	Mukhopadhyay et al. [6], Venkateshwaran and Elayaperumal [10]
Vakka	N/A	Density ~810 kg/m^3 Tensile strength ~549 MPa Elastic modulus ~15.85 GPa Elongation ~3.46%	Rao and Rao [8]
Date palm (spadix stem)	Cellulose (%) ~43.05 Hemicellulose (%) ~27.48 Lignin (%) ~29.47 (Sukkari cultivar)	Density ~1090–1370 kg/m^3 Tensile strength ~33–680 MPa Elastic modulus ~8.81–15 GPa Elongation ~3.33% Thermal conductivity ~0.087 W/mK (chemical retting)	Elseify et al. [13]
Thespesia Lampas	Cellulose (%) ~60.63 Hemicellulose (%) ~26.64 Lignin (%) ~12.70	Tensile strength ~573 MPa Elastic modulus ~61.2 GPa Elongation ~0.79%	Reddy et al. [16]

(continued)

Table 4.1 (continued)

Fiber	Composition	Properties	Refs.
Okra	Cellulose (%) ~67.5 Hemicellulose (%) ~15.4 Lignin (%) ~7.1	Tensile strength ~281 MPa Elastic modulus ~16.55 GPa	Alam and Khan [19]
Hibiscus sabdariffa (bleached)	Cellulose (%) ~58 Hemicellulose (%) ~8.15 Lignin (%) ~7.9	Density ~1490 kg/m^3 Bundle strength ~15.6 g/tex Moisture ~8.76% Elongation ~14.5%	Kalita et al. [21]
Bamboo	Cellulose (%) ~45–50 Hemicellulose (%) ~20–25 Lignin (%) ~20–30	Density ~600–1100 kg/m^3 Tensile strength ~140–230 MPa Elastic modulus ~17.2 GPa Elongation ~1.4–1.7%	Ramirez et al. [23]
Bagasse	Cellulose (%) ~50 ± 5 Hemicellulose (%) ~22.5 ± 2.5 Lignin (%) ~21 ± 3	Density ~720–880 kg/m^3 Elastic modulus ~15–19 GPa Elongation ~1–5%	Yadav and Gupta [36]
Alfa fibers	Cellulose (%) ~46 ± 3 Lignin (%) ~20 ± 2	Density ~1400 kg/m^3 Tensile strength ~229 MPa Elastic modulus ~18.42 GPa Elongation ~1.5%	Helaili and Chafra [37]
Mendong fibers	Cellulose (%) ~72.14 Hemicellulose (%) ~20.2 Lignin (%) ~3.44	Density ~892 kg/m^3 Tensile strength ~452 ± 47 MPa Elastic modulus ~17.4 ± 3.9 GPa Elongation ~	Suryanto et al. [33]
Napier grass	Cellulose (%) ~45 Hemicellulose (%) ~34 Lignin (%) ~20.5	Density ~892 kg/m^3 Tensile strength ~75 MPa Elastic modulus ~6.8 GPa Elongation ~2.8%	Reddy et al. [34]

4.5 Conclusions

The conventional plant fibers (hemp, jute, sisal, etc.) are produced in less quantity and have relatively higher cost. The non-conventional fibers can serve as a better replacement for these fibers. These plants are abundantly available, and researchers have explored their potential to serve as a source of fibers. These fibers may be used for conventional textiles in certain cases, but also have a huge potential to serve as a reinforcement material for composites. It will provide the composite designers and engineers a wide range of options in terms of biodegradable reinforcements. Focusing more on the non-conventional sources of natural fibers will help to maintain a cleaner and greener environment, thus contributing to environment sustainability.

References

1. N.Z. Zakaria, M.Z. Sulieman, R. Talib, Turning natural fiber reinforced cement composite as innovative alternative sustainable construction material: a review paper. Int. J. Adv. Eng. Manag. Sci. 1(8), 24–31 (2015)
2. O. Güven, S.N. Monteiro, E.A.B. Moura, J.W. Drelich, Re-emerging field of lignocellulosic fiber—polymer composites and ionizing radiation technology in their formulation. Polym. Rev. 56(4), 702–736 (2016). https://doi.org/10.1080/15583724.2016.1176037
3. L. Mathew, K.U. Joseph, R. Joseph, Isora fibre: morphology, chemical composition, surface modification, physical, mechanical and thermal properties—A potential natural reinforcement. J. Nat. Fibers 3(4), 13–27 (2007). https://doi.org/10.1300/j395v03n04_02
4. C.J. Chirayil, J. Joy, L. Mathew, M. Mozetic, J. Koetz, S. Thomas, Isolation and characterization of cellulose nanofibrils from Helicteres isora plant. Ind. Crops Prod. 59, 27–34 (2014). https://doi.org/10.1016/j.indcrop.2014.04.020
5. J. Joy, C. Jose, S.B. Varanasi, L.P. Mathew, S. Thomas, S. Pilla, Preparation and characterization of poly(butylene succinate) bionanocomposites reinforced with cellulose nanofiber extracted from Helicteres isora plant. J. Renew. Mater. 4(5), 351–364 (2016). https://doi.org/10.7569/JRM.2016.634128
6. S. Mukhopadhyay, R. Fangueiro, Y. Arpaç, Ü. Şentürk, Banana fibers—Variability and fracture behaviour. J. Eng. Fiber. Fabr. 3(2), 39–45 (2008)
7. K. Ebisike, B. AttahDaniel, B. Babatope, S. Olusunle, Studies on the extraction of naturally-occurring banana fibers. Int. J. Eng. Sci. 2(9), 9 (2013)
8. K.M.M. Rao, K.M. Rao, Extraction and tensile properties of natural fibers: vakka, date and bamboo. Compos. Struct. 77(3), 288–295 (2007). https://doi.org/10.1016/j.compstruct.2005.07.023
9. J.L. Guimarães, F. Wypych, C.K. Saul, L.P. Ramos, K.G. Satyanarayana, Studies of the processing and characterization of corn starch and its composites with banana and sugarcane fibers from Brazil. Carbohydr. Polym. 80(1), 130–138 (2010). https://doi.org/10.1016/j.carbpol.2009.11.002
10. N. Venkateshwaran, A. Elayaperumal, Banana fiber reinforced polymer composites—A review. J. Reinf. Plast. Compos. 29(15), 2387–2396 (2010). https://doi.org/10.1177/0731684409360578
11. K. Murali Mohan Rao, K. Mohana Rao, A.V. Ratna Prasad, Fabrication and testing of natural fibre composites: vakka, sisal, bamboo and banana, in *Materials and Design*, vol. 31, no. 1 (Elsevier Ltd, 2010), pp. 508–513
12. K. Ramanaiah, A.V.R. Prasad, K.H.C. Reddy, Thermophysical and fire properties of vakka natural fiber reinforced polyester composites. J. Reinf. Plast. Compos. 32(15), 1092–1098 (2013). https://doi.org/10.1177/0731684413486366
13. L.A. Elseify, M. Midani, L.A. Shihata, H. El-Mously, Review on cellulosic fibers extracted from date palms (Phoenix Dactylifera L.) and their applications. Cellulose 26(4), 2209–2232 (2019). https://doi.org/10.1007/s10570-019-02259-6
14. E.A. Elbadry, Agro-residues : surface treatment and characterization of date palm tree fiber as composite reinforcement, vol. 2014 (2014)
15. F.M. AL-Oqla, O.Y. Alothman, M. Jawaid, S.M. Sapuan, M.H. Es-Saheb, Processing and properties of date palm fibers and its composites, in *Biomass and Bioenergy: processing and Properties*, ed. by K.R. Hakeem, M. Jawaid, U. Rashid (Springer, Berlin 2014), pp. 1–67
16. K.O. Reddy, B. Ashok, K.R.N. Reddy, Y.E. Feng, J. Zhang, A.V. Rajulu, Extraction and characterization of novel lignocellulosic fibers from Thespesia lampas plant. Int. J. Polym. Anal. Charact. 19(1), 48–61 (2014). https://doi.org/10.1080/1023666X.2014.854520
17. F. Anwar, R. Qadir, N. Ahmad, Cold pressed okra (Abelmoschus esculentus) seed oil, in *Cold Pressed Oils* (Elsevier, 2020), pp. 309–314

18. I.M. De Rosa, J.M. Kenny, D. Puglia, C. Santulli, F. Sarasini, Morphological, thermal and mechanical characterization of okra (Abelmoschus esculentus) fibres as potential reinforcement in polymer composites. Compos. Sci. Technol. **70**(1), 116–122 (2010). https://doi.org/10.1016/j.compscitech.2009.09.013

19. M.S. Alam, G.M.A. Khan, Chemical analysis of okra bast fiber (Abelmoschus esculentus) and its physico-chemical properties, J. Text. Apparel Technol. Manag. **5**(4) (2007)

20. A.S. Singha et al., Surface-modified hibiscus sabdariffa fibers: physicochemical, thermal, and morphological properties evaluation. Int. J. Polym. Anal. Charact. **14**(8), 695–711 (2009). https://doi.org/10.1080/10236660903325518

21. B.B. Kalita, S. Jose, S. Baruah, S. Kalita, S.R. Saikia, Hibiscus sabdariffa (Roselle): a potential source of bast fiber. J. Nat. Fibers **16**(1), 49–57 (2019). https://doi.org/10.1080/15440478.2017.1401504

22. A.P. Deshpande, M. Bhaskar Rao, C. Lakshmana Rao, Extraction of bamboo fibers and their use as reinforcement in polymeric composites. J. Appl. Polym. Sci. **76**(1), 83–92 (2000). https://doi.org/10.1002/(SICI)1097-4628(20000404)76:1<83::AID-APP11>3.0.CO;2-L

23. F. Ramirez, A. Maldonado, J. Correal, M. Estrada, Bamboo-Guadua angustifolia kunt fibers for green composites, in *18th International Conference on Composite Matter*, pp. 1–4 (2011)

24. X. U. Wei, T. Ren-Cheng, Extracting natural bamboo fibers from crude bamboo fibers by caustic treatment. Biomass Chem. Eng.**40**(3), 1–5 (2006)

25. S. Jain, R. Kumar, U.C. Jindal, Mechanical behaviour of bamboo and bamboo composite. J. Mater. Sci. **27**(17), 4598–4604 (1992). https://doi.org/10.1007/BF01165993

26. Use of Sugarcane Bagasse Fibre in Apparel Industry. *Fibre2Fashion* (2013). https://www.fibre2fashion.com/industry-article/7048/sugarcane-fibres. Accessed 17 Apr 2021

27. D. Michel, B. Bachelier, J.-Y. Drean, O. Harzallah, Preparation of cellulosic fibers from sugarcane for textile use, in *Conference Papers in Science*, vol. 2013, pp. 1–6 (2013). https://doi.org/10.1155/2013/651787

28. W. Hoareau, F.B. Oliveira, S. Grelier, B. Siegmund, E. Frollini, A. Castellan, Fiberboards based on sugarcane bagasse lignin and fibers. Macromol. Mater. Eng. **291**(7), 829–839 (2006). https://doi.org/10.1002/mame.200600004

29. S.M. Luz, A.R. Gonçalves, A.P. Del'Arco, Mechanical behavior and microstructural analysis of sugarcane bagasse fibers reinforced polypropylene composites. Compos. Part A Appl. Sci. Manuf. **38**(6), 1455–1461 (2007). https://doi.org/10.1016/j.compositesa.2007.01.014

30. J. Sun, R. Sun, X. Sun, Y. Su, Fractional and physico-chemical characterization of hemicelluloses from ultrasonic irradiated sugarcane bagasse. Carbohydr. Res. **339**(2), 291–300 (2004). https://doi.org/10.1016/j.carres.2003.10.027

31. A. Maghchiche, A. Haouam, B. Immirzi, Extraction and characterization of algerian alfa grass short fibers (stipa tenacissima). Chem. Chem. Technol. **7**(3), 339–344 (2013)

32. A. Ben Mabrouk, H. Kaddami, S. Boufi, F. Erchiqui, A. Dufresne, Cellulosic nanoparticles from alfa fibers (Stipa tenacissima): extraction procedures and reinforcement potential in polymer nanocomposites. Cellulose**19**(3), 843–853 (2012). https://doi.org/10.1007/s10570-012-9662-z

33. H. Suryanto, E. Marsyahyo, Y.S. Irawan, R. Soenoko, Morphology, structure, and mechanical properties of natural cellulose fiber from mendong grass (Fimbristylis globulosa). J. Nat. Fibers **11**(4), 333–351 (2014). https://doi.org/10.1080/15440478.2013.879087

34. K.O. Reddy, C.U. Maheswari, M. Shukla, A.V. Rajulu, Chemical composition and structural characterization of Napier grass fibers. Mater. Lett.**67**(1), 35–38 (2012). https://doi.org/10.1016/j.matlet.2011.09.027

35. K.O. Reddy, C.U. Maheswari, D.J.P. Reddy, A.V. Rajulu, Thermal properties of Napier grass fibers. Mater. Lett.**63**(27), 2390–2392 (2009). https://doi.org/10.1016/j.matlet.2009.08.035

36. S. Yadav, G. Gupta, R. Bhatnagar, A review on composition and properties of bagasse fibers (2015)

37. S. Helaili, M. Chafra, Anisotropic visco-elastic properties identification of a natural biodegradable ALFA fiber composite. J. Compos. Mater. **48**(13), 1645–1658 (2014). https://doi.org/10.1177/0021998313488811

Chapter 5
Wood and Agriculture Waste Fibers

Abstract The lignocellulosic fibers are obtained from leaf, fruit, seed, bast, grass, agrowaste and wood. This chapter is focused on the fibers obtained from wood and agriculture waste. The wood fibers have a vast potential for being used as reinforcement to get composites for non-structural applications. The agriculture waste fibers are either obtained from yard trimmings, husk, bagasse or bark of the plant. Traditionally majority of agrowaste is either burnt or land filled. Focus on fiber extraction from agrowaste may be regarded a step towards circularity and sustainability. All these fiber categories are discussed in this chapter. The fiber chemical composition and some specific properties are also compared for these fibers.

Keywords Wood fibers · Husk and straw fibers · Fibers from bark of plant

5.1 Wood Fibers

The lignocellulosic fibers may be broadly classified into two categories depending on the origin, i.e., from wood and from annual plants [1]. Both these categories have specific technologies and supply chain to produce the fibers. The pulp mills process the wood fibers for paper and boards [2], while plant fibers are processed to produce yarns and fabrics. Contrary to the plant fibers having small lumen area (0–5%), the wood fibers possess a high luminal area (20–70%). More the size of lumen, lower is the strength and stiffness of the fibers. Additionally, the wood fibers have low cellulose crystallinity than plant fibers (55–70% and 90–95%, respectively). Chemical treatment of the wood fiber helps to lower the moisture absorption process, clean and modify the fiber surface to increase the surface roughness [3]. In wood fibers, the microfibrillar angle ranges 3–50° depending on the fiber location in the wood, while the microfibrillar angle in plant fibers ranges 6–10°. Typically, the wood fibers are short in length (1–5 mm) and may be used to fabricate composites with random fiber orientation. Such composites have isotropic in-plane properties. The chemical composition of some common wood fibers is given in Table 5.1.

Two distinct approaches are used to fabricate composite reinforced with wood fiber. In case of thermosetting resin, impregnating wood fiber mat with resin gives composite material. Commingling technique is preferred for thermoplastic matrices,

K. Shaker and Y. Nawab, *Lignocellulosic Fibers*,
SpringerBriefs in Materials, https://doi.org/10.1007/978-3-030-97413-8_5

Table 5.1 Composition of wood fibers from different sources [4]

	Cellulose (w%)	Hemicellulose (w%)	Lignin (w%)
Spruce	49	20	29
Balsa	48	28	22
Cedar	44	21	30
Pine	42	29	28
Birch	41	32	22
Poplar	39	28	30

i.e., production of mats having wood pulp fibers and thermoplastic fibers. These commingled mats are then compression molded to produce the desired composite parts [4]. The wood fiber composites are mainly used for the automotive applications. These composites are either prepared by compression or injection molding technique [5]. The random fiber distribution results in moderate mechanical properties, and are therefore preferred for non-structural components like door/boot liners, parcel shelves, etc.

5.2 Husk and Straw Fibers

5.2.1 Areca Fruit Husk

Areca is a monocotyledonous plant and belonging to the palm species, and fibers are extracted from the husk of Areca fruit. Binoj et al. [6] studied the chemical, physical, mechanical, and thermal properties of Areca Fruit Husk (AFH) fibers. The Areca nuts are used for tobacco industry, and husk covering the nuts are generally discarded. For fiber extraction, the husks are soaked in water for five days, and rinsed 5–10 times every day with running water. The so-obtained fibers were dried in sun for 3 days and brushed to extract the fiber strands. The fiber composition was found to have high cellulose (57.35–58.21%), hemicellulose (13–15.42%) and lignin content (23.17–24.16%). It is a light weight fiber having low density (0.78 g/cm^3). The moisture content in these fibers was 7.32%, almost similar to the other lignocellulosic fibers. The fibers exhibit a strength of 147, 196, 259 and 322 MPa during tensile test for gauge length of 10, 20, 30 and 40 mm respectively. The modulus varied from 1.12 to 3.15 GPa, with a failure strain of 10–13%. The thermogravimetric analysis shows that fiber is stable up to 240 °C, and therefore can be used for reinforcing polymers having processing temperature below this temperature.

Binoj et al. [7] fabricated composite materials using unsaturated polyester as resin and AFH fibers as reinforcement material. The tensile test showed these composites to have a strength, modulus and elongation value of 68.2 MPa, 1.32 GPa and 5.16% respectively at 40% fiber loading (optimum value). Similarly, the values obtained in

Table 5.2 Chemical composition of rice husk

Component	Cellulose (%)	Hemicelluloses (%)	Lignin (%)	Crude protein (%)	Ashes (%)
Fraction	35	25	20	2	17

flexural test were 73.91 and 1.62 for strength and modulus respectively. The impact strength of these composites was 6.82 J/cm^2.

Srinivasa et al. [8] used AFH fibers as reinforcement and produced composites with epoxy resin. They showed that the flexural and impact properties are a function of fiber loading and tend to increase with higher fiber percentage. They treated the fibers with KOH solution to modify its surface properties, leading to the increased fiber-matrix adhesion. Thus, the chemical treatment affected the mechanical performance, and the flexural and impact strength of treated areca fiber composites were increased in comparison to the untreated fiber composites.

5.2.2 Rice Husk

The rice husk is a hard protective covering of rice, naturally tough, water insoluble, having abrasive resistance behavior. The exterior surface of rice husk is silica coated with a thick cuticle. Small amounts of silica are present in the mid region and inner epidermis of rice husk. It is used as filler for the composite materials, and its chemical composition is given in Table 5.2. The researchers have focused the physical and mechanical properties of composites obtained using different type of matrix materials to obtain composites like polyethylene, polypropylene, thermosets, etc. as matrix material [9]. A number of efforts are underway to further enhance the performance of composites by applying certain chemical treatments to rice husk fibers. Another approach used is the fabrication of hybrid composites with rice husk as one component, in addition to matrix and fibrous reinforcement.

Suhot et al. [10] reviewed the developments in rice husk reinforced polymer composites between 2017 and 2021. The researchers have focused on the use of RH as fillers, their modification and treatment, and developed hybrid composites. Another aspect studied in the recent publications was moisture absorption of RH loaded composites, composite swelling and their diffusion coefficients. It was also reported that the addition of rice husk has enhanced the flame retardancy of the developed polymer composites.

5.2.3 Corn Husk

Corn (maize) is one of the major commercial agriculture crops. The different agriculture waste from this crop includes husk, cob, silk and stalk. Green leaves outside.

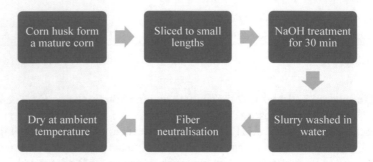

Fig. 5.1 Fiber extraction process from corn husk

Cob is the central core of the corn, the outside green leaves covering is husk, long shiny fibers at the top of a corn ear are called corn silk. The fiber extraction process from corn husk is shown in Fig. 5.1.

The dry fiber obtained were further treated with alkali or enzyme. The alkali treated cornhusk fibers had 64.5% cellulose and 6.4% lignin. Huda and Yang [11] used these fibers with polypropylene fibers to fabricate composite materials by hot compression. The resultant composites showed better sound absorption and higher tensile and flexural properties as compared to the jute/PP composites.

5.2.4   Corn Stalk/Straw

Corn is a major food crop and its bye product, i.e., corn stalk is used as a source of fibers. Conventionally, mechanical, and chemical methods are used to get fibers from corn stalk. The mechanical means included steam, ammonia and carbon dioxide explosion, while chemical methods include alkaline treatment. Both these approaches produce fibers that are not suitable for textile or composite material applications. Reddy and Yang [15] used an optimized process for the extraction of high quality fibers and studied their structure and properties. They treated the pithed corn stalk with 2% NaOH solution at 95 °C for 45 min in a closed container. The slurry was then washed and filtered to get fibers; that were subsequently neutralized, and dried. The cellulose and lignin in these fibers was 81% and 8.4% respectively.

5.2.5   Wheat Straw

Sain and Panthapulakkal [12] used a two stage process to extract the fibers from wheat straw. First stage was mechanical defibrillation of straw and next was the fungal retting. To facilitate the mechanical action, straws were soaked overnight in water. The retted fibers were stronger than un-retted fibers, having lower fiber length and diameter. The FTIR analysis showed removal of hemicellulose and lignin in the retted fibers. The thermal degradation analysis showed stability of upto 250 °C. Reddy and Yang [13] prepared long cellulosic fibers from wheat straw and characterized their performance. They used a detergent pretreatment and mechanical action, followed by alkaline treatment to get high quality fibers. The wheat straw is high resistant to alkalies, and pretreatment with detergent helps to remove this wax layer. These fibers had an average length of 40–80 mm, with breaking strength comparable to cotton and kenaf fibers.

5.2.6   Soybean Straw

Soybean is one of the most important oil crop in the world. The soybean straw, a byeproduct, can serve as a source of fibers. Reddy and Yang [14] investigated a number of treatment conditions to extract fibers from soybean straw. The optimized condition was selected on the basis of fiber fineness and yield % obtained. The soybean straws were boiled in a 8% NaOH for 2 h, thoroughly washed and neutralized with acetic acid. The fibers were allowed to dry under ambient conditions, and had a length to width ratio of approx 1000:1. The cellulose and lignin in these fibers was found to be 85 ± 3.1% and 11.8 ± 0.7% respectively. The fiber fineness was 76 denier, with a tenacity of 2.7 g/den, nearly equal to that of the cotton fiber, although the elongation of this fiber is half to that of the cotton.

Moisture regain in these fibers was found to be 7.9%, almost equal to that of the cotton fiber (8.5%). The elongation in cornstalk fibers is 1.1–3.5%, almost similar to that of flax fiber (2–3%). The strength of these fibers shows two distinct ranges, i.e., 1.5–2.5 g/den and 3.5 g/den. The lower strength range is similar to kenaf, while high strength is similar to that of cotton fiber. Li et al. [16] used fibers from corn straw as natural oil sorbents. The acetylated corn straw fibers are oleophilic and do not get wet with water. The maize tassel is also used as a source of cellulose fibers [17]. A similar procedure is adopted for the fiber extraction, as discussed for corn stalk.

5.3   Bark of Plant

5.3.1   Acacia Planifrons

Acacia Planifrons is a small plant having umbrella-like spreading branches and grows up to 7 m heigh. Fibers were separated from the bark of this plant by immersing in water for a period of 14 days [18]. The fibers are extracted mechanically using ultrafine long metal teeth. The extracted fibers are then washed and allowed to dry for 4 days at ambient temperature. The cellulose, hemicellulose and lignin were estimated to be 73.1%, 9.41%, and 12.04% respectively in these fibers. The density of these fibers was found to be 0.66 g/cm^3. The thermogravimetric analysis of these fibers exhibited their higher thermal stability of up to 270 °C.

5.3.2 Cotton Stalk

Cotton is one of the most widely used fiber for textile applications. It is widely cultivated across the globe and its production was 24.42 million metric tons in 2020–21. The USA, China, India, Brazil and Pakistan are the largest producers of cotton fiber. The fibers are conventionally obtained from the seed of this plant. However, the stalks of cotton plant, left in the fields is also a source of lignocellulosic fibers. The fibers obtained from cotton stalk had far superior mechanical properties as compared to the other lignocellulosic fibers obtained from agriculture waste like wheat straw or rice husk.

For fiber extraction, the outer bark of stalk is stripped off, air dried and cut into small (8 cm) segments. Dong et al. [19] extracted long fine fibers (27 dtex) from the bark of cotton stalks by steam explosion process, following KOH and peroxide treatment. The fibers had a smooth and clean surface, giving high carding yield of 68.6%. The chemical composition of extracted fibers was found to be majorly cellulose (82.1%), lignin (11.8%) and hemicellulose (4.4%). The high cellulose content is due to the removal of lignin and hemicellulose due to KOH treatment. The fibers were used as reinforcement for composite materials but could not be spun into yarn.

5.3.3 Dichrostachys Cinerea

The bark of Dichrostachys Cinerea is green and hairy on young branches, and grey-brown on older branches and stems. Baskaran et al. [20] extracted fibers from the mature and fresh bark of Dichrostachys Cinerea plant. To extract fibers, the barks were submerged in water for 15 days, peeled and washed in running water, and dried at ambient temperature. Finally, the fibers were separated by mechanical scrapping using metal comb like structure. The analysis of extracted fibers revealed that it has 72.4% cellulose, 13.1% hemicellulose and 16.9% lignin. The moisture content is 9.8% and density of fiber was determined to be 1240 kg/m^3. The fibers showed a tensile strength of 873 ± 14 MPa. In the thermogravimetric analysis, the fibers were found to be thermally stable up to 226 °C. The fibers showed a mass loss of 13.7% at 285 °C, probably due to depolymerization of hemicellulose.

In another study, Baskaran et al. [21] investigated the effect of alkali-treatment on the performance of Dichrostachys Cinerea fibers and their composites. The treated fibers had a relatively lower hemicellulose percentage (4.6%) and increased crystallinity index. Its moisture content was also reduced to 6.65%. Contrary to this, the tensile properties were significantly enhanced from 820 to 885 MPa by alkali-treatment. These outcomes showed the potential of DCF to be used as reinforcement for green composites. The composites fabricated with epoxy resin showed a tensile strength and modulus of 47.64 and 490 MPa respectively, at a fiber volume fraction of 40%. Impact strength of these composites was 76.55 J/m

5.3.4   Prosopis Juliflora

Fibers from the bark of Prosopis juliflora plant were first extracted and characterized by Saravanakumar et al. [22]. This plant has flexible branches and can grow in wide range of soils, and mostly found in arid and semiarid regions. Bark of this plant is a source of lignocellulosic fibers. For fiber extraction, the plant was immersed in water for 14 days to allow the microbial degradation. The outer layer was disposed while fibers were separated from inner layer by combing. The structural analysis of isolated fibers showed its composition to be 61.65% cellulose, 16.14% hemicellulose and 17.11% lignin. The moisture content of these fibers was 9.48%, with a remarkably low density of 580 kg/m^3.

They treated the extracted fibers with a NaOH [23], and found that hemicellulose content was significantly reduced. The composition of alkali treated fibers was 68.6% cellulose, 6.1% hemicellulose and 14.2% lignin. The density of treated fibers was increased to 648 kg/m^3. The extracted fibers were subjected to a number of other chemical treatments including benzoyl peroxide, potassium permanganate and stearic acid, and their properties were investigated [24]. The stearic acid treated fibers showed highest cellulose content of 82.16%, and highest density of 933 kg/m^3. Similarly, the stearic acid treated fibers were found to be more thermally stable, as compared to other treatments. Researchers [25, 26] have used these fibers as reinforcement for composites and investigated their performance properties.

5.4   Other Agrowaste Fibers

This section includes some other agrowaste sources that have been investigated for the lignocellulosic fibers, as shown in Fig. 5.2.

The Grewia Optiva is a plant found in the Himalayan region of South Asia and is used to feed animals. The locals used to extract fibers from the shoots of this plant by water retting and subsequent crushing by stones. They use the fiber to make articles like ropes, floor mats, bags, coats, etc. These fibers have a density of 1.2 g/cm^3 and their chemical composition comprises of 55–65% cellulose, 20–25% hemicellulose and 3–4% lignin [27]. The tensile testing showed that the fibers have a breaking strength of 15.35 MPa at 10.28% elongation. Researchers have used various methods such as delignification, mercerization, benzoyl, silane treatment, and graft copolymerization to improve the efficiency of Grewia Optiva fibers. Singha and Thakur [28] evaluated the properties of untreated and silane-treated Grewia Optiva fibers. It has been widely explored by the researchers as a source of lignocellulosic reinforcement

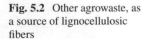

Fig. 5.2 Other agrowaste, as a source of lignocellulosic fibers

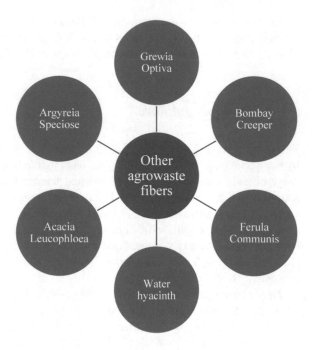

for the composite materials. Rana et al. [27] have reviewed the chemical treatments of these fibers and the matrices used by different researchers to fabricate composites of these fibers.

Bombay Creeper (Vernonia elaeagnifolia), also called as curtain creeper is an ornamental plant. Shaker et al. [29] extracted fibers from the yard trimmings of Bombay creeper by water retting. The cut stems were immersed in water for 14 days, and the fibers were extracted from degraded stems. The extracted fibers were oven dried and their physical, thermal and mechanical properties were characterized. The fiber fineness was found to be 22.25 tex, which shows that these are quite coarse fibers. The density and moisture content of the fibers was 1.36–1.6 g/cm^3 and 9.17%, comparable to other lignocellulosic fibers. The tenacity of these fibers was found to be 16–20 cN/tex, which is almost equal to that of the cotton fiber.

Ferula communis is a perennial plant with 1–2.5 m height and is also known as Chakshir. Seki et al. [30] extracted fibers from F. communis, identified its microstructure, thermal behavior, chemical composition and mechanical properties. The cut stems of plant were immersed in boiling acetone (20% aq. solution) for 30 min and were then left in the solution for 48 h. After due time, the fibers were extracted. The ferula communis fibers contain cellulose (53.3%), hemicellulose (8.5%), lignin (1.4%), ash content (7.0%) and other compounds (29.8%). Crystallinity index of these fibers (48%) is close to that of kapok fiber (46%) and less than that of jute, sisal and hemp (71%, 71% and 88% respectively). The ferula fibers have tensile strength, tensile modulus and breaking elongation of 475.6 MPa, 52.7 GPa and 4.2% respectively. These properties are comparable to the tensile properties of jute, ramie and pineapple fiber, and higher than coir, cotton and luffa fibers. They also fabricated composite materials by impregnating these fibers with unsaturated polyester resin, and fiber–matrix interfacial shear strength in these composites was found to be 16.21 MPa.

Water hyacinth (Scientific Name: Eichhornia crassipes) is an aquatic plant, floating freely on the water surface. Fibers extracted from these plants have high cellulose content and small diameter (25–50 μm) and are therefore valued to be used as reinforcement for composites [31]. For fiber extraction, the chopped hyacinth plant was dewaxed and bleached two times. The hemicellulose was removed by treating it with sodium hydroxide solution, while lignin was removed by sodium chlorite treatment. The samples were centrifuged and washed with distilled water. Abral et al. [32] fabricated composite materials using unsaturated polyester resin and hyacinth fibers, with varying fiber volume fractions. The fibers were also treated with various alkali concentrations, and their effect on mechanical properties was investigated. Another variable of the study was immersion in water. It was found that the tensile and flexural properties of composite material reduced after immersion in water. The optimum mechanical performance was observed at a fiber loading of 15%.

Arthanarieswaran et al. [33] extracted fibers from bark of Acacia leucophloea plant by degumming processes. The plants were submerged in water for a period of two weeks for microbial degradation. Subsequently, the fibers were segregated from the bark, washed tap water and dried for one week. The fibers were treated with 5% NaOH solution at varying process parameters, neutralized, dried and stored for testing. Density of these fibers was found to be 1.38–1.43 g/cm^3. In thermogravimetric analysis, the onset of degradation peak was observed at 220 °C, showing degradation of hemicellulose. The fibers with optimum treatment showed a chemical composition of cellulose (76.6%), hemicellulose (3.8%) and lignin (13.6%).

Fibers extracted form Argyreia Speciose are a potential reinforcement for composite materials. The fiber is locally termed as "Tezgam" owing to its high growth rate. Shaker et al. [34] reported the extraction of fibers from this creeper plant by water retting and their mechanical properties and thermal stability. The linear density of these fibers was found to be 11.2 tex and moisture regain of 10%. The tenacity of untreated and alkali treated fibers was 4.61 and 5.91 cN/tex. These fibers were found to be thermally stable up to a temperature of 250 °C. They also reported the use of these fibers as filler to reinforce the composite materials [35]. These fillers significantly enhanced the mechanical properties of developed composites. The tensile properties were increased by approximately 30%. The addition of fillers also increased the brittleness of the developed composite materials.

References

1. H.P.S.A. Khalil, M.S. Alwani, A.K.M. Omar, Chemical composition, anatomy, lignin distribution, and cell wall structure of Malaysian plant waste fibers. BioResources **1**, 220–232 (2006). https://doi.org/10.15376/biores.1.2.220-232
2. G.V. Duarte, B.V. Ramarao, T.E. Amidon, P.T. Ferreira, Effect of hot water extraction on hardwood kraft pulp fibers (acer saccharum, sugar maple). Ind. Eng. Chem. Res. **50**(17), 9949–9959 (2011). https://doi.org/10.1021/ie200639u
3. A.S. Singha, V.K. Thakur, Morphological, thermal, and Physicochemical characterization of surface modified Pinus fibers. Int. J. Polym. Anal. Charact. **14**(3), 271–289 (2009). https://doi.org/10.1080/10236660802666160
4. B. Madsen, E.K. Gamstedt, Wood versus plant fibers : similarities and differences in composite applications, vol. 2013 (2013)
5. K.G. Satyanarayana, F. Wypych, J.L. Guimarães, S.C. Amico, T.H.D. Sydenstricker, L.P. Ramos, Studies on natural fibers of Brazil and green composites. Met. Mater. Process. **17**(3–4), 183–194 (2005)
6. J.S. Binoj, R.E. Raj, V.S. Sreenivasan, G.R. Thusnavis, Morphological, physical, mechanical, chemical and thermal characterization of sustainable Indian Areca Fruit Husk Fibers (Areca Catechu L.) as potential alternate for hazardous synthetic fibers. J. Bionic Eng. **13**(1), 156–165 (2016). https://doi.org/10.1016/S1672-6529(14)60170-0
7. J.S. Binoj, R.E. Raj, B.S.S. Daniel, S.S. Saravanakumar, Optimization of short Indian Areca fruit husk fiber (Areca catechu L.)—Reinforced polymer composites for maximizing mechanical properties. Int. J. Polym. Anal. Charact. **21**(2), 112–122 (2016). https://doi.org/10.1080/1023666X.2016.1110765
8. C.V. Srinivasa et al., Static bending and impact behaviour of areca fibers composites. Mater. Des. **32**(4), 2469–2475 (2011). https://doi.org/10.1016/j.matdes.2010.11.020
9. N. Bisht, P.C. Gope, N. Rani, Rice husk as a fibre in composites: a review. J. Mech. Behav. Mater. **29**(1), 147–162 (2020). https://doi.org/10.1515/JMBM-2020-0015/MACHINEREADA BLECITATION/RIS
10. M.A. Suhot, M.Z. Hassan, S.A. Aziz, M.Y. Md Daud, Recent progress of rice husk reinforced polymer composites: a review. Polymers (Basel) **13**(15) (2021). https://doi.org/10.3390/pol ym13152391
11. S. Huda, Y. Yang, Chemically extracted cornhusk fibers as reinforcement in light-weight poly(propylene) composites. Macromol. Mater. Eng. **293**(3), 235–243 (2008). https://doi.org/10.1002/mame.200700317
12. M. Sain, S. Panthapulakkal, Bioprocess preparation of wheat straw fibers and their characterization. Ind. Crops Prod. **23**(1), 1–8 (2006). https://doi.org/10.1016/j.indcrop.2005.01.006
13. N. Reddy, Y. Yang, Preparation and characterization of long natural cellulose fibers from wheat straw. J. Agric. Food Chem. **55**(21), 8570–8575 (2007). https://doi.org/10.1021/jf071470g
14. N. Reddy, Y. Yang, Natural cellulose fibers from soybean straw. Bioresour. Technol. **100**(14), 3593–3598 (2009). https://doi.org/10.1016/j.biortech.2008.09.063
15. N. Reddy, Y. Yang, Structure and properties of high quality natural cellulose fibers from cornstalks. Polymer (Guildf) **46**(15), 5494–5500 (2005). https://doi.org/10.1016/j.polymer.2005.04.073
16. D. Li, F.Z. Zhu, J.Y. Li, P. Na, N. Wang, Preparation and characterization of cellulose fibers from corn straw as natural oil sorbents. Ind. Eng. Chem. Res. **52**(1), 516–524 (2013). https://doi.org/10.1021/ie302288k
17. C.E. Maepa, J. Jayaramudu, J.O. Okonkwo, S.S. Ray, E.R. Sadiku, J. Ramontja, Extraction and characterization of natural cellulose fibers from Maize Tassel. Int. J. Polym. Anal. Charact. **20**(2), 99–109 (2015). https://doi.org/10.1080/1023666X.2014.961118
18. P. Senthamaraikannan, S.S. Saravanakumar, V.P. Arthanarieswaran, P. Sugumaran, Physicochemical properties of new cellulosic fibers from the bark of Acacia planifrons. Int. J. Polym. Anal. Charact. **21**(3), 207–213 (2016). https://doi.org/10.1080/1023666X.2016.1133138

19. Z. Dong, X. Hou, F. Sun, L. Zhang, Y. Yang, Textile grade long natural cellulose fibers from bark of cotton stalks using steam explosion as a pretreatment. Cellulose **21**(5), 3851–3860 (2014). https://doi.org/10.1007/s10570-014-0401-5

20. P.G. Thirukumaran, M. Ramesh, P. Sivaramakrishnan, D.U. Baravanakumar, Characterization of new natural cellulosic fiber from the Bark of Dichrostachys Cinerea. J. Nat. Fibers **15**(1), 62–68 (2018). https://doi.org/10.1080/15440478.2017.1304314

21. P.G. Baskaran, M. Kathiresan, P. Pandiarajan, Effect of alkali-treatment on structural, thermal, tensile properties of dichrostachys cinerea bark fiber and its composites. J. Nat. Fibers **0478** (2020). https://doi.org/10.1080/15440478.2020.1745123

22. S.S. Saravanakumar, A. Kumaravel, T. Nagarajan, P. Sudhakar, R. Baskaran, Characterization of a novel natural cellulosic fiber from Prosopis juliflora bark. Carbohydr. Polym. **92**(2), 1928–1933 (2013). https://doi.org/10.1016/j.carbpol.2012.11.064

23. S.S. Saravanakumar, A. Kumaravel, T. Nagarajan, I.G. Moorthy, Investigation of physico-chemical properties of alkali-treated prosopis juliflora fibers. Int. J. Polym. Anal. Charact. **19**(4), 309–317 (2014). https://doi.org/10.1080/1023666X.2014.902527

24. S.S. Saravanakumar, A. Kumaravel, T. Nagarajan, I.G. Moorthy, Effect of chemical treatments on physicochemical properties of Prosopis juliflora fibers. Int. J. Polym. Anal. Charact. **19**(5), 383–390 (2014). https://doi.org/10.1080/1023666X.2014.903585

25. P.V. Reddy, D. Mohana Krishnudu, P. Rajendra Prasad, V.S.R.R, A study on alkali treatment influence on Prosopis Juliflora fiber-reinforced epoxy composites. J. Nat. Fibers **18**(8), 1094–1106 (2021). https://doi.org/10.1080/15440478.2019.1687063

26. B. Surya Rajan, M.A.S. Balaji, S.S. Saravanakumar, Effect of chemical treatment and fiber loading on physico-mechanical properties of Prosopis juliflora fiber reinforced hybrid friction composite. Mater. Res. Express **6**(3) (2019). https://doi.org/10.1088/2053-1591/aaf3cf

27. U. Ahmed, Y. Nawab, M. Umair, A. Raza, Development of light weight and flexible neck guard fabric for ice Hockey players, in *4th International Textile Conference* (2020)

28. A.S. Singha, V.K. Thakur, Synthesis and characterizations of silane treated Grewia optiva fibers. Int. J. Polym. Anal. Charact. **14**(4), 301–321 (2009). https://doi.org/10.1080/102366 60902871470

29. K. Shaker et al., Extraction and characterization of novel fibers from Vernonia elaeagnifolia as a potential textile fiber, Ind. Crop. Prod. **152**, 112518 (2020). https://doi.org/10.1016/j.ind crop.2020.112518

30. Y. Seki, M. Sarikanat, K. Sever, C. Durmuşkahya, Extraction and properties of Ferula communis (chakshir) fibers as novel reinforcement for composites materials. Compos. Part B Eng. **44**(1), 517–523 (2013). https://doi.org/10.1016/j.compositesb.2012.03.013

31. M. Thiripura Sundari, A. Ramesh, Isolation and characterization of cellulose nanofibers from the aquatic weed water hyacinth—Eichhornia crassipes. Carbohydr. Polym. **87**(2), 1701–1705 (2012). https://doi.org/10.1016/j.carbpol.2011.09.076

32. H. Abral et al., Mechanical properties of water hyacinth fibers—Polyester composites before and after immersion in water. Mater. Des. **58**, 125–129 (2014). https://doi.org/10.1016/j.mat des.2014.01.043

33. V.P. Arthanarieswaran, A. Kumaravel, S.S. Saravanakumar, Physico-chemical properties of Alkali-Treated Acacia leucophloea fibers. Int. J. Polym. Anal. Charact. **20**(8), 704–713 (2015). https://doi.org/10.1080/1023666X.2015.1081133

34. K. Shaker, Y. Nawab, Extraction and characterization of fiber from Argyreia Speciosa creeper plant, in *International Conference on Sugar Palm and Allied Fibre Polymer Composites 2021*, pp. 48–50 (2021)

35. K. Shaker, M. Umair, S. Shahid, M. Jabbar, Cellulosic fillers extracted from argyreia speciose waste : a potential reinforcement for composites to enhance properties. J. Nat. Fibers (2020)

Chapter 6
Performance of Green Composites

Abstract The lignocellulosic fiber reinforced composite materials exhibit comparable specific properties to those reinforced with synthetic fibers. These composites
are advantageous in terms of their biodegradability, annual renewability and environmental friendliness. However, the green composites are not suitable to replace
their synthetic counterparts due to some noteworthy limitations including moisture
absorption, high flammability, thermal degradation, inconsistent fiber properties,
etc. This chapter focusses on the mechanical properties of some green composites, reinforced with novel lignocellulosic fibers. The limitations of these composites
(moisture absorption, thermal stability, flammability and biodegradation) and their
potential remedies are discussed later in the chapter.

Keywords Mechanical properties · Moisture absorption · Thermal degradation ·
Biodegradability

6.1 Mechanical Performance

The mechanical properties of a composite material are of utmost importance to
decide about their suitability for a particular application area. The mechanical performance of a composite is governed by the reinforcing material. The lignocellulosic
fibers exhibit comparable specific stiffness and specific strength as compared to
glass fibers [1]. However, it must be considered that due to the bio-based origin of
the natural fibers, their properties like fiber length, fineness, stiffness, etc. show a
remarkable variation. The variation in performance properties of fiber is attributed
to the factors like location, seeding density, climate, soil quality, and harvesting [2],
which also determines the chemical composition of fiber. Additionally, the fiber-
matrix adhesion, fiber orientation, dispersion and fiber volume fraction also influence of the mechanical behavior of the composites [3]. A better fiber orientation
and longer fiber tend to improve the tensile strength of yarn/composite. Researchers
have also explored the performance of hybrid composites having both lignocellulosic and synthetic fibers used as reinforcement. It is done to increase the mechanical
performance of the composites [4], as green composites have relatively low mechanical properties. Table 6.1 summarizes the properties of novel lignocellulosic fiber

© The Author(s), under exclusive license to Springer Nature Switzerland AG 2022   57
K. Shaker and Y. Nawab, *Lignocellulosic Fibers*,
SpringerBriefs in Materials, https://doi.org/10.1007/978-3-030-97413-8_6

Table 6.1 Properties of novel fiber reinforced composites

Fibrous reinforcement	Matrix	Fiber loading (wt%) (%)	Fabrication technique	Properties				Refs.
				Tensile Strength (MPa)	Tensile Modulus (GPa)	Elongation at break (%)	Others	
Agave	PLA	30	Compression molding	35	2030	1.9	–	Cisneros-López et al. [5]
Agave	PHB	30	Milled fiber/polymer blending and subsequent compression molding	15.7	770	–	FM ~2154 MPa FS ~20.5 MPa	Torres-Tello et al. [6]
Agave	PHBV	30	Milled fiber/polymer blending and subsequent compression molding	15.4	673	–	FM ~2370 MPa FS ~17.2 MPa	
Curaua	PHB (triethyl citrate used as plasticizer)	10	Milled fiber/polymer blending, and then injection molding	21	220	10	IS ~67 J/m TC at 40 °C ~0.1647 W/(m K)	Scalioni et al. [7]
Curaua mercerized	PHB (triethyl citrate used as plasticizer)	10	Milled fiber/polymer blending, and then injection molding	25	178	14	IS ~70 J/m TC at 40 °C ~0.2313 W/(m K)	

(continued)

Table 6.1 (continued)

Fibrous reinforcement	Matrix	Fiber loading (wt%) (%)	Fabrication technique	Properties			Others	Refs.
				Tensile Strength (MPa)	Tensile Modulus (GPa)	Elongation at break (%)		
Curaua fiber acetone treated	PHB (triethyl citrate used as plasticizer)	10	Milled fiber/polymer blending, and then injection molding	19	158	14	IS ~71 J/m TC at 40 °C ~0.1864 W/(m K)	
Hemp	PBAT	20	Injection molding	6.9	370	3.7	Toughness ~18.4 J/m³	Pulikkalakath Ambil et al. [8]
Hemp, silane treated	PBAT	20	Injection molding	8.8	425	3.6	Toughness ~22.6 J/m³	
Kenaf	PHBV	40	Fiber mixing in molten polymer, and subsequent press molding	–	–	–	FM ~3.6 GPa FS ~16.6 MPa IS ~2.2 kJ/m²	Wang and Mao [9]
		60		–	–	–	FM ~4.02 GPa FS ~14.8 MPa IS ~1.7 kJ/m²	
Kenaf	PBAT	40	Fiber mixing in molten polymer, and subsequent press molding	–	–	–	FM ~713 MPa FS ~9.61 MPa IS ~9.6 kJ/m²	
		60		–	–	–	FM ~939 MPa FS ~6.01 MPa IS ~6.0 kJ/m²	

(continued)

Table 6.1 (continued)

| Fibrous reinforcement | Matrix | Fiber loading (wt%) (%) | Fabrication technique | Properties | | | | Refs. |
				Tensile Strength (MPa)	Tensile Modulus (GPa)	Elongation at break (%)	Others	
Kenaf + Coir	PLA	30 + 30	Fiber soaking in PLA suspension, drying and hot pressing	110	7200	–	FM ~14.5 GPa FS ~172 MPa	Yusoff et al. [10]
Bamboo + Coir	PLA	30 + 30	Fiber soaking in PLA suspension, drying and hot pressing	154	5900	–	FM ~9 GPa FS ~210 MPa	
Kenaf + Bamboo + Coir	PLA	20 + 20 + 20	Fiber soaking in PLA suspension, drying and hot pressing	185	7300	–	FM ~8.9 MPa FS ~205 MPa	

Where, IS—Impact Strength, TC, Thermal Conductivity, FS—Flexural Strength and FM—Flexural Modulus

Table 6.2 Tensile properties of wood fiber composites [12]

	Fiber architecture	Fiber volume fraction (%)	Strength (MPa)	Modulus (GPa)
Wood pulp/polypropylene	RD	27	28	4.2
Eucalyptus saw dust/unsaturated Polyester	RD	46	60	6.2
Kraft + TMP fibers/polypropylene	RD	40	43	4.5
Sulphite pulp/polypropylene	RD	50	51	3.9
Kraft paper/phenol formaldehyde	RD	72	247 (MD)	26.2 (MD)
Kraft paper/phenol formaldehyde	RD	72	156 (CD)	11.7 (CD)

Where, RD—Random distribution, TMP—Thermomechanical pulp, MD—Machine direction and CD—Cross direction

reinforced composite materials.

The wood fiber composites are mainly used for the automotive applications [11]. These composites are either prepared by injection molding or compression molding techniques. The random fiber distribution results in moderate mechanical properties, and are therefore preferred for non-structural components like door liners, boot liners, parcel shelves, etc. The properties of wood and plant fiber composites are compared in Table 6.2.

6.2 Moisture Absorption

The chemical composition and structure of lignocellulosic fibers has been discussed in Chap. 2. The hydroxyl groups present in the cellulose structure tend to attract water and form hydrogen bonding. It results in the absorption of moisture by the fiber [13]. The moisture absorption behavior of fibers is advantageous in a number of applications but for reinforcement purposes, it is highly undesirable. The moisture affects the interfacial bonding between fiber and matrix, thus creating regions of poor load transfer and deteriorating its mechanical properties. The capillary effect transports water through the interface, which causes the reduction in interfacial adhesion. Leads to a decrease in the fiber-matrix adhesion, creating regions of poor load transfer, reducing the mechanical properties of composite, deformation, and even failure [14]. In case of green composites, the issue becomes more critical as bio-based matrices tend to absorb higher amount of moisture than their petroleum-based

Table 6.3 Moisture absorption % of some lignocellulosic fibers

Fiber	Moisture absorption (%)
Kapok	9.86
Brazil nut	Bur fibers ~14.73 Shell fibers ~10.29
Agave Americana	8
Pineapple leaf fiber	11.8
Corn stalk	7.9
Dichrostachys Cinerea	9.8
Banana	10–11
Bamboo	12

counterparts. The other parameters contributing to moisture absorption are the relative humidity and temperature of the environment and the amount of fibers present in the composite material. The fiber surface is composed of non-cellulosic substances like lignin, pectin and waxes, that prevent adequate fiber-matrix adhesion. Both these issues are addressed by the modification of natural fiber surfaces using some physical or chemical treatments [15]. Moisture absorption % of some lignocellulosic fibers is shown in Table 6.3.

The researchers have adopted a number of approaches to minimize the moisture absorption of lignocellulosic fibers, to enhance the fiber-matrix interfacial adhesion and improve the mechanical properties. Ali et al. [13] reviewed the different approaches used including mercerization, acetylation, silane, enzyme, plasma, sodium chlorite, per-oxide, ozone treatment, etc. They concluded that selecting a suitable technique will help to reduce moisture absorption and fibers may be used effectively to reinforce the green composites.

6.3   Flammability and Thermal Degradation

The ability of a material to burn, resulting in its combustion is termed as flammability. It is also defined as the degree of difficulty to cause combustion of the material [16]. The lignocellulosic fibers have inherently poor flame retardancy and are therefore more susceptible to catch flame during any fire hazard. In the form of composite material, the matrix covers the reinforcement, providing some protection against fire. Hence the polymer matrix defines the flame retardant behavior of the resulting composite material, instead of reinforcement. But the polymeric materials are also highly flammable, regardless of the origin (bio- or petroleum-based) [17]. The fiber/polymer burning results in the release of heat energy and toxic combustion products (soot, particles, ashes, etc.). The test methods commonly used to test flame retardancy include ignition time, heat release rate, flame spread, smoke production and toxicity.

Contrary to flammability, thermal stability defines the ability of a material to resist decomposition under thermal stress, and to maintain its properties (strength, elasticity, toughness, etc.) at given temperature. The thermal stability of a fiber/polymer is generally determined by the thermogravimetric analyzer, in terms of weight loss %. The thermal stability depends on the chemical structure of polymer, molecular weight and degree of crystallinity. The aromatic structures present in polymer chain help to improve its thermal stability. Contrary to this, the double bonds or oxygen-containing structures in the backbone make the polymer less resistant to high temperatures [17]. Therefore, it is desirable to introduce cyclic, heterocyclic systems, aromatic rings, ester, amide, carbonate, sulphone and disulfide groups preferably in the main chain of polymer at an early stage during synthesis. The alternative way to impart thermal stability and flame retardancy is by the addition of flame retardant fillers in the matrix and/or treatment of reinforcement. Both these approaches delay the ignition or retard the burning process, and/or suppressing the formation of smoke [18].

There are certain compounds (metal hydroxides and carbonates) that absorb heat delivered to the polymer and decompose endo-thermally by dehydration or decarboxylation. The release of water molecules further delays the burning process and reduces its temperature. Such compounds can also be used as filler material to induce flame retardancy. A typical example is hydrated aluminum oxide or aluminum hydroxide. When ignited, it decomposes into aluminum oxide, releasing water, and forms a protective layer on the polymer surface to delay the flame spread. The magnesium hydroxide also behaves in a similar way, but it has higher thermal stability than aluminum oxide (the decomposition temperature of 320–350 °C and 200–230 °C, respectively).

6.4 Biodegradation

The biodegradable polymers have been under research since 1980s and were initially produced by mixing conventional polymers with starch. The decomposition of starch resulted in disintegration of the structure into small fragments of conventional polymer. But such polymers cannot be regarded as true biodegradable polymers but were only bio disintegrated. The ASTM D6400 defines the degradable plastics as "a plastic in which degradation results from the action of naturally occurring microorganisms such as bacteria, fungi, and algae" [19].

The sustainability issues have fostered the development of biodegradable polymers from renewable resources, e.g., plants. Some of these issues include cost of petroleum based raw material, renewability issues, global warming, post-consumer pollution, etc. Ideally, the plant-based biodegradable polymers will decompose to CO_2 and water after a short exposure to the weather. Being plant origin, these polymers will be CO_2 neutral [20]. The biodegradation of green composite materials is determined by the degradation of its constituents. The lignocellulosic fibers are inherently biodegradable. The polar groups in cellulose and hemicellulose provide a moist environment, that favors the growth of microorganisms, inducing biodegradation.

Using degradable bio-based polymers will result into a completely biodegradable product [21].

The degradation of green composites is affected by the constituents, environmental conditions (moisture, temperature, microorganism activity, UV radiation, etc.). These factors collectively impact the fiber/matrix interfacial adhesion, causing micro-cracks in the composite, accelerating its biodegradation. The growth of microorganisms on green composites also leads to deterioration of mechanical properties such as tensile and flexural strength, hardness, etc. The test methods for biodegradability study of polymers include gravimetry, respirometry, surface hydrolysis, radiolabeling, GPC/SEC, GC/MS, AFM, TEM, NMR, SEM, FTIR, etc. These techniques have been compared by Ewa Rudnik in terms of polymeric form being degraded, inoculum and degradation criteria [19].

6.5 Conclusions

The growing issues of global warming, plastic pollution and depletion of petroleum resources have forced the greening of world, leading towards sustainable developments. The green composites have emerged as a sustainable class of composite materials, to replace the conventional synthetic counterparts. The book presented lignocellulosic fibers obtained from some novel sources as a potential replacement of synthetic reinforcements. Some of the fibers explored are abundantly available, and are not from edible resources, creating no issue for food chain. The composition and properties of these fibers are also compared in the chapters, allowing the selection of fibers most suitable for a certain end use. The researchers have used these fibers as reinforcement with bio-based polymers. Some examples are mentioned in the Sect. 6.1. The moisture absorption, flammability and thermal stability of these fibers are least explored by researchers. There is a huge potential to explore the approaches for reducing the moisture absorption and flammability of these fiber-based composites. The approaches used for conventional synthetic composites may be used, but it still needs to be investigated.

References

1. K. Shaker, Y. Nawab, M. Jabbar, Bio-composites : eco-friendly substitute of glass fiber composites, in *Handbook of Nanomaterials and Nanocomposites for Energy and Environmental Applications*, ed. by O.V. Kharissova, L.M.T. Martínez, B.I. Kharisov (Springer, 2020)
2. M.P.M. Dicker, P.F. Duckworth, A.B. Baker, G. Francois, M.K. Hazzard, P.M. Weaver, Green composites: a review of material attributes and complementary applications. Compos. Part A Appl. Sci. Manuf. **56**, 280–289 (2014). https://doi.org/10.1016/j.compositesa.2013.10.014
3. K.L. Pickering, M.G.A. Efendy, T.M. Le, A review of recent developments in natural fibre composites and their mechanical performance. Compos. Part A Appl. Sci. Manuf. **83**, 98–112 (2016). https://doi.org/10.1016/j.compositesa.2015.08.038

4. M.F. Ismail, M.T.H. Sultan, A. Hamdan, A.U.M. Shah, A study on the low velocity impact response of hybrid Kenaf-Kevlar composite laminates through drop test rig technique. BioResources **13**(2), 3045–3060 (2018). https://doi.org/10.15376/biores.13.2.3045-3060
5. E.O. Cisneros-López et al., Polylactic acid resin/nat ural fiber composite produced by rotational molding: a comparative study with compression molding. Adv. Polym. Technol. **37**(7), 2528–2540 (2018). https://doi.org/10.1002/adv.21928
6. E.V. Torres-Tello, J.R. Robledo-Ortíz, Y. González-García, A.A. Pérez-Fonseca, C.F. Jasso-Gastinel, E. Mendizábal, Effect of agave fiber content in the thermal and mechanical properties of green composites based on polyhydroxybutyrate or poly(hydroxybutyrate-co-hydroxyvalerate). Ind. Crops Prod.**99**, 117–125 (2017). https://doi.org/10.1016/j.indcrop.2017.01.035
7. L.V. Scalioni, M.C. Gutiérrez, M.I. Felisberti, Green composites of poly(3-hydroxybutyrate) and curaua fibers: morphology and physical, thermal, and mechanical properties. J. Appl. Polym. Sci. **134**(14), 44676 (2017). https://doi.org/10.1002/APP.44676
8. H. Pulikkalparambil, J. Parameswaranpillai, J.J. George, K. Yorseng, S. Siengchin, Physical and thermo-mechanical properties of bionano reinforced poly(butylene adipate-co-terephthalate), hemp/CNF/Ag-NPs composites. AIMS Mater. Sci. **4**(3), 814–831 (2017). https://doi.org/10.3934/matersci.2017.3.814
9. J. Wang, Q. Mao, Biodegradable composites reinforced with kenaf fibers: thermal, mechanical, and morphological issues. Adv. Polym. Technol. **32**(2013), 474–485 (2012). https://doi.org/10.1002/adv
10. R.B. Yusoff, H. Takagi, A.N. Nakagaito, Tensile and flexural properties of polylactic acid-based hybrid green composites reinforced by kenaf, bamboo and coir fibers. Ind. Crops Prod. **94**, 562–573 (2016). https://doi.org/10.1016/j.indcrop.2016.09.017
11. K.G. Satyanarayana, F. Wypych, J.L. Guimarães, S.C. Amico, T.H.D. Sydenstricker, L.P. Ramos, Studies on natural fibers of Brazil and green composites. Met. Mater. Process. **17**(3–4), 183–194 (2005)
12. B. Madsen, E.K. Gamstedt, Wood versus plant fibers : similarities and differences in composite applications, vol. 2013 (2013)
13. A. Ali et al., Hydrophobic treatment of natural fibers and their composites—A review. J. Ind. Text. **47**(8), 2153–2183 (2018). https://doi.org/10.1177/1528083716654468
14. T. Väisänen, O. Das, L. Tomppo, A review on new bio-based constituents for natural fiber-polymer composites. J. Clean. Prod. **149**, 582–596 (2017). https://doi.org/10.1016/j.jclepro.2017.02.132
15. A.Y. Al-Maharma, N. Al-Huniti, Critical review of the parameters affecting the effectiveness of moisture absorption treatments used for natural composites. J. Compos. Sci.**3**(1) (2019). https://doi.org/10.3390/jcs3010027
16. M. Fan, F. (Structural engineer) Fu, *Advanced High Strength Natural Fibre Composites in Construction* (Woodhead Publishing, 2017)
17. A. Sienkiewicz, P. Czub, Flame retardancy of biobased composites—Research development. Materials (Basel) **13**(22), 1–30 (2020). https://doi.org/10.3390/ma13225253
18. A. Shrivastava, *Introduction to Plastics Engineering* (Elsevier, 2018)
19. K. Król-Morkisz, K. Pielichowska, Thermal decomposition of polymer nanocomposites with functionalized nanoparticles, in *Polymer Composites with Functionalized Nanoparticles: synthesis, Properties, and Applications*, ed. by K. Pielichowski, T.M. Majka (Elsevier Inc., 2018), pp. 405–435
20. E. Vázquez-Núñez, A.M. Avecilla-Ramírez, B. Vergara-Porras, M.R. del López-Cuellar, Green composites and their contribution toward sustainability: a review. Polym. Polym. Compos.**29**(9), S1588–S1608 (2021). https://doi.org/10.1177/09673911211009372
21. A. Czapla, M. Ganesapillai, J. Drewnowski, Composite as a material of the future in the era of green deal implementation strategies. Processes**9**(2238) (2021)

Printed in the United States
by Baker & Taylor Publisher Services